Passport to Intermediate Algebra

Michael R. Anderson
Department of Mathematics
West Virginia State University

KENDALL/HUNT PUBLISHING COMPANY
4050 Westmark Drive   Dubuque, Iowa 52002

Cover image copyright © Photodisc

Copyright © 2005 by Michael Anderson

ISBN 0-7575-1660-2

Kendall/Hunt Publishing Company has the exclusive rights to reproduce this work,
to prepare derivative works from this work, to publicly distribute this work,
to publicly perform this work and to publicly display this work.

All rights reserved. No part of this publication may be reproduced,
stored in a retrieval system, or transmitted, in any form or by any
means, electronic, mechanical, photocopying, recording, or otherwise,
without the prior written permission of Kendall/Hunt Publishing Company.

Printed in the United States of America
10 9 8 7 6 5 4 3 2 1

Contents

Introduction   vii

1 Basic Stuff   1
   1.1 Sets 1
   1.2 Arithmetic 3
   1.3 Inequalities and Absolute Value 4
   1.4 Exponents 6

2 Equations and Inequalities   9
   2.1 Solving Equations 9
   2.2 Formulas 11
   2.3 Story Problems 12
   2.4 Inequalities 15
   2.5 Absolute Value Equations and Inequalities 17
   2.6 Do-It-Yourself Formulas 20

3 The Coordinate Plane and Functions   23
   3.1 Ordered Pairs 23
   3.2 The Rectangular Coordinate System 24
   3.3 What is a Function? 26
   3.4 Reading a Graph 28

4 All About Lines   31
   4.1 Drawing Lines 31
   4.2 The Slope of a Line 34
   4.3 Equations of Lines 37
   4.4 Solving Things with Graphs 39
   4.5 More Things to Solve with Graphs 40

4.6 Using Graphs 42

5 More Than One Equation — 45
5.1 Two Lines at the Same Time 45
5.2 Story Problems and Systems of Equations 49
5.3 Three Equations in Three Variables 52
5.4 Linear Inequalities in Two Variables 55
5.5 Systems of Linear Inequalities in Two Variables 58

6 Polynomials — 61
6.1 Polynomial ABCs 61
6.2 Combining Functions 63
6.3 Adding and Subtracting Polynomials 64
6.4 Multiplying Polynomials 64
6.5 Starting to Factor 67
6.6 Special Factoring Formulas 68
6.7 Factoring Trinomials 69
6.8 Dividing Polynomials 71
6.9 Solving Equations by Factoring 74

7 Rational Expressions — 77
7.1 Simplifying Rational Expressions 77
7.2 Multiplying and Dividing Rational Expressions 80
7.3 Adding and Subtracting Rational Expressions 82
7.4 Complex Fractions 85
7.5 Solving Rational Equations 87
7.6 Story Problems and Rational Equations 89
7.7 Negative Exponents 92

8 Radical Expressions — 95
8.1 ABCs of Radicals 95
8.2 Simplifying Radicals 98
8.3 Combining Radicals 100
8.4 Solving Radical Equations 103
8.5 The Pythagorean Theorem 106
8.6 Fractional Exponents 107
8.7 Complex Numbers 109

CONTENTS

9 Quadratic Things — 113
- 9.1 Graphing a Quadratic Function . . . 113
- 9.2 Completing the Square . . . 115
- 9.3 The Quadratic Formula . . . 118
- 9.4 More Equations . . . 121
- 9.5 Quadratic Inequalities . . . 123

10 Conic Sections — 127
- 10.1 Graphing Parabolas . . . 127
- 10.2 Circles . . . 131
- 10.3 Ellipses and Hyperbolas . . . 132
- 10.4 Nonlinear Systems . . . 136

11 Exponentials and Logarithms — 139
- 11.1 Inverse Functions . . . 139
- 11.2 Exponentials . . . 141
- 11.3 Logarithms . . . 144
- 11.4 Properties of Logarithms . . . 146
- 11.5 Logarithmic and Exponential Equations . . . 148

Introduction

What This Book Isn't

This book is not a substitute for your textbook or your teacher, nor will it help you if you don't do math problems. Sorry, you still have to buy the textbook, go to class and do your homework.

What This Book Is

This book is a supplement to your regular intermediate algebra textbook. To be specific, it's a supplement to *Intermediate Algebra*, 4th edition, by Streeter, Hutchison, Bergman and Hoelzle(published by McGraw Hill). Intermediate algebra, however, is standardized enough that it will probably work as a supplement to most other intermediate algebra texts, too.

How to Use This Book

The purpose of this book is to help you learn and retain the ideas covered in your course. I've tried to do this by highlighting the important ideas in each section and by presenting the material in a less formal and technical manner than your text. Nonetheless, you should have your regular textbook handy while you're reading this book, both to get the details of a particular concept and to start learning how to read a mathematics textbook.

On the other hand, keep this book handy while you're using your textbook, as a quick reference for the important ideas in the section you're working on, and, perhaps, as the source for a friendlier explanation of the material.

Reading Your Textbook

Mathematics textbooks tend to be written in a style all of their own. The language tends to be very compressed – every word counts. (Mathematicians are not paid by the word.) Furthermore, the language is somewhat technical – words have special meanings. This can lead to confusion when the reader sees a word and expects it to have its ordinary meaning while the author expects the reader to use the mathematical meaning. An example of this is the word "evaluate". *Webster's New Collegiate Dictionary* gives as its first definition "to determine or fix the value of" and its second as "to determine the significance or worth of usually by careful appraisal and study" and gives "estimate" as a synonym. The first definition is close to what a math textbook means by "evaluate" (usually something along the lines of "replace all occurrences of a letter in an expression by the given number and then do the operations"), but definitely not the second definition, which seems to be the most common use of the word now. (I have an old dictionary.) And as for the synonym "estimate"? In non-mathematical settings, it is a synonym, but in math, it's not.

Another standard feature of math books is all those symbols. In math textbooks, symbols are just abbreviations. As mentioned above, mathematicians are not paid by the word, so they would much rather write "$3x+2$" than "three times some quantity plus two". It certainly takes up less space. So get to know the symbols. Here are the most common ones in algebra books:

- Operations: $+$, $-$, $\times$, $\div$, $\sqrt{}$

- Variables: x, y, z

- Constants: a, b, c

- Functions: f, g, h

Note that multiplication is often indicated by writing the factors next to each other or by a dot ($\cdot$), and division is often indicated by a slash (/) or a fraction bar.

Thirdly, it helps to look at the structure of a section in your textbook. If we randomly consider Section 5.3 in Streeter/Hutchison/Bergman/Hoelzle, it seems to be built like this:

1. Definition (or concept)

2. Worked example

3. Student exercise similar to example

4. Worked example (a variation of the first one)

5. Student exercise similar to second example

6. General concept or method

7. Worked example

8. Student exercise similar to third example

9. Worked example

10. Student exercise similar to fourth example

11. Worked applied example (story problem)

12. Student exercise similar to fifth example

13. Worked applied example (story problem)

14. Student exercise similar to sixth example

The authors of the book actually expect you to work through those exercises in the text. And that brings me to the final point: Reading a math textbook must be done *actively* rather than *passively*. By passive reading, I mean you look at the words and read them and that's it. In active reading, you'll still read the words, but you'll also get out your paper and pencil and work through the examples and exercises while you are reading. Ask any mathematician and you'll be told that the only way to learn math is by doing it, and the best time to do it is while you're reading your textbook.

So to summarize:

1. Pay attention to how the author uses words.

2. Learn the symbols.

3. Keep the structure of the section in mind.

4. Read actively.

Acknowledgments

First, I would like to thank all of the students I've had in all of my years of teaching. I've probably learned more from you than you have from me, and without you, I probably wouldn't have written this book. Second, I'd like to thank two people at Kendall/Hunt Publishing: Eric Klosterman, for suggesting that I write this book, and Emily Cabbage, my Development Editor, for helping me through the process of creating and publishing this book. Finally, I would like to thank Jame McCumbee and Tonda Sue Tryon, who read and commented on this book while it was still in manuscript. Of course, any errors that remain are entirely my responsibility. If you find any, contact me at *andersmr@mail.wvstateu.edu* and if there are future editions of this book, they'll be corrected there.

Chapter 1

Basic Stuff

You've probably seen all of the ideas in Chapter 1 already. But look it over anyway to make sure you haven't missed something, and to pick up the notation which will be used throughout the rest of the book.

1.1 Sets

What's Important

- How to write sets
- The different kinds of numbers
- Number lines

Section 1.1 starts off with sets. A set is just a collection of things, so you can talk about a set of dishes or a set of students, or, in a math class, a set of numbers.

In mathematics, one way to write down a set is to list the members (the technical term is **elements**), like this:

$$\{\text{Fred}, \text{Barney}\}$$

or like this:

$$\{1, 2, 3, 4, 5\}$$

or even like this:

$$\{1, 2, 3, 4, \ldots\}$$

This method is good for small sets.

A second way is to find a phrase that describes everything in the set and use what is called *set-builder notation*. For example, the last set above could be written this way in set-builder notation:

$$\{x | x \text{ is a natural number}\}$$

This method is good for sets where the members have something in common.

Some other information you may need about sets:

- The order you write the elements of a set doesn't matter.

- Two sets are equal if they contain exactly the same things (naturally).

- Sets can be *finite* (see the first two sets) or *infinite* (the third set).

Mathematicians give names to seven different sets of numbers:

1. The **natural numbers** (denoted by $\mathbb{N}$), sometimes called the **counting numbers**: $\{1, 2, 3, \ldots\}$.

2. The **whole numbers**: $\{0, 1, 2, 3, \ldots\}$.

3. The **integers** (denoted by $\mathbb{Z}$). Think of these as the positive and negative whole numbers: $\{\ldots, -2, -1, 0, 1, 2, \ldots\}$.

4. The **rational numbers** (denoted by $\mathbb{Q}$), usually called **fractions**. If you write one as a decimal, it eventually stops (a *terminating decimal*) or the digits start repeating (a *repeating decimal*). Examples of rational numbers would be:

$$\frac{5}{8}, -\frac{157}{487204}, 7.38, 5.2\overline{43}$$

5. The **irrational numbers**, which are all the decimals that don't terminate and don't repeat. $\sqrt{2}$ and π are standard examples of irrational numbers.

6. The **real numbers** (denoted by $\mathbb{R}$). To make the real numbers, take all the numbers I've listed above and stuff them into one set.

7. The **complex numbers** (denoted by $\mathbb{C}$). You'll look at the complex numbers in Section 8.7.

1.2. ARITHMETIC

Most of the examples and homework problems are written using integers and fractions (or decimals), but you may use the real numbers for answers, or even the complex numbers for some equations.

Real numbers can be pictured by drawing a number line. A typical way to label the number line is to put 0 in the middle, then go $1, 2, 3$ and so on off to the right and $-1, -2, -3$ and so on off to the left, which gives you a picture like this:

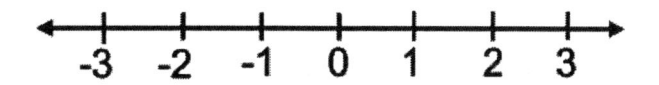

1.2 Arithmetic

What's Important

- Order of operations

This section talks about addition, subtraction, multiplication and division. I'll assume you already know how to add, subtract, multiply and divide real numbers, but I do want to remind you of a few things: First, the *order of operations*.

1. Do the operations in parentheses (brackets, braces) first.

2. Then do exponents.

3. Then multiply and divide.

4. Finally, add and subtract.

A handy acronym to remember is **PEMDAS**, which stands for
  Parentheses
  Exponents
  Multiplication &
  Division
  Addition &
  Subtraction

Example: $(4 + 8) \div 6 + 2 = 12 \div 6 + 2 = 2 + 2 = 4$

Second, some important properties of addition and multiplication are the *commutative*, *associative* and *distributive* properties. The commutative property says that the order you add or multiply things doesn't matter, you'll get the same thing. The associative property says that it doesn't matter how you group things when adding or multiplying, you'll get the same thing. The distributive property describes how addition and multiplication interact:

$$a(b+c) = ab + ac$$

Example: $3(x+y) = 3x + 3y$

The distributive law is actually being used when you combine like terms. Remember that like terms have the some combinations of letters (variables) raised to the same powers. You combine them by adding or subtracting the numbers in front (the *coefficients*).

Example: $4x + 3y + 8x^2 + 5y = 8x^2 + 4x + 8y$ when the like terms ($3y$ and $5y$) have been combined. Note that since the $4x$ and the $8x^2$ have different powers on the x, they're not like terms and so you can't combine them together.

1.3 Inequalities and Absolute Value

What's Important

- Writing inequalities
- Graphing inequalities
- Absolute value

Inequalities (**greater than, less than**, etc.) are really about relative position on the number line. If one number is to the left of a second number, then we say the first number is *less than* the second number. If the first number is to the right of the second number, then we say the first number is *greater than* the second number. For example, 212 is to the left of 483 so we write $212 < 483$.

In algebra, we often have inequalities involving variables, e.g., $x > 6$. A whole bunch of numbers will make this inequality true, such as 7 and 23 and 35 and .... In fact, all numbers greater than 6 will be solution of this inequality. There are two standard ways to write down the solution set. First, you can use set notation:

$$\{x | x > 6\}$$

1.3. INEQUALITIES AND ABSOLUTE VALUE

Second, you can graph it on a number line:

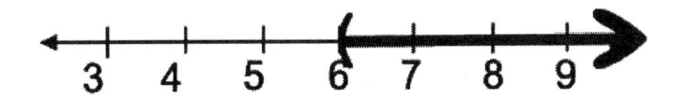

Some people prefer to use circles rather than parentheses or brackets:

More complicated inequalities can be written, too. For example, we can have this inequality:

$$73 < x \leq 145$$

An inequality like this is built out of the two separate inequalities $x > 73$ and $x \leq 145$ joined together. The solution to this is all numbers greater than 73 *and* less than or equal to 145. (Notice I keep saying "and". That's an important part of this inequality.) The graph of the solution looks like:

Or in the circle style:

Another idea we can get from position on the number line is **absolute value** (notation: $|a|$). The absolute value of a number is its distance from zero along the number line. For example, 28 is 28 units away from 0, so $|28| = 28$. And -482 is 482 units away from 0, so $|-482| = 482$.

The algebraic definition for absolute value looks like this:

$$|a| = \begin{cases} a & \text{if } x \geq 0 \\ -a & \text{if } x < 0 \end{cases}$$

If you check the examples from above using this definition, we still have $|28| = 28$ since $28 \geq 0$ and we still have $|-482| = 482$ since $-482 < 0$ and $-(-482) = 482$.

1.4 Exponents

> What's Important
> - Working with exponents
> - Scientific notation

Natural number exponents are a convenient way to represent repeated multiplication:

$$\underbrace{a \cdot a \cdot a \cdots a}_{n \text{ factors of } a} = a^n.$$

(Terminology: a is called the **base**, n is called the **exponent**.) So, for example, $3^4 = 3 \cdot 3 \cdot 3 \cdot 3 = 81$.

We can also use 0 as an exponent:

$$a^0 = 1 \text{ if } a \neq 0.$$

(We don't define a value for 0^0. If you actually try this on a calculator, you'll probably get an error message.) Hence, for example, $(-147230)^0 = 1$.

Later on, we'll also look at using negative integers, rational numbers and real numbers as exponents.

There are five important rules for working with exponents:

1. $a^m \cdot a^n = a^{m+n}$. If you multiply two expressions that have the same base, then add their exponents.

2. $\dfrac{a^m}{a^n} = a^{m-n}$. If you divide two expressions that have the same base, then subtract the bottom exponent from the top exponent.

3. $(a^m)^n = a^{mn}$. If you raise an exponent to another exponent, then multiply the exponents.

4. $(ab)^n = a^n b^n$. If you raise a product to an exponent, then raise each factor to that exponent.

5. $\left(\dfrac{a}{b}\right)^n = \dfrac{a^n}{b^n}$. If you raise a quotient to an exponent, then raise the top to the exponent and raise the bottom to the exponent.

1.4. EXPONENTS

Warning: Notice that we didn't say anything about raising sums or differences to exponents. For those situations, you'll have to go back to the original definition of exponents and distribute things out.

The rules above are used in simplifying expressions involving exponents.

Example:

$$(3x^2)^2 \cdot (2xy^3)^4 = 3^2(x^2)^2 \cdot 2^4 x^4 (y^3)^4$$
$$= 9x^4 \cdot 16x^4 y^{12}$$
$$= 144 x^8 y^{12}$$

One application of exponents is called **scientific notation**. It's a way to compactly write very large numbers and very small numbers. The basic form:

$$a \times 10^n,$$

where a is a number between 1 and 10 ($1 \leq a < 10$ and n is an integer. (For the moment, n will just be a positive integer.) To put a number into scientific notation, count how many places you would have to move the decimal point to make the number between 1 and 10. That will be your n value.

Example: $57,100,000,000,000 = 5.71 \times 10^{13}$

To go the other way, move the decimal point to the right n places.

Example: $4.83 \times 10^6 = 4,830,000$

By converting large numbers to scientific notation, you can use the exponent rules to multiply or divide the numbers.

Example:

$$\frac{57,300,000,000}{3,000,000} = \frac{5.73 \times 10^{10}}{3 \times 10^6}$$
$$= \frac{5.73}{3} \times \frac{10^{10}}{10^6}$$
$$= 1.91 \times 10^4$$
$$= 19,100$$

Chapter 2

Equations and Inequalities

In this chapter we look at how to solve linear equations (Sections 2.1–2.3), how to solve linear inequalities of various kinds (Sections 2.4-2.5) and how to build your own formulas (Section 2.6).

2.1 Solving Equations

What's Important

- How to solve a linear equation.

Linear equations are equations that can be written like this:

$$ax + b = 0$$

where a and b are constants.

Example: $4x + 8 = 0$

Example: $5x - 2 = 3x + 8$

Note that the second example doesn't look like the form in the definition, but by rearranging things and combining like terms, we can put it in that form. The way this is usually said is: The equations are *equivalent*.

Solving a linear equation involves rearranging things and combining like terms until your equation looks like $x =$ some number. To actually do the rearranging, you use these properties:

- you can add or subtract the same number from both sides and the answer won't change;

- you can multiply or divide by the same nonzero number on both sides and the answer won't change.

Example: Solve $4x + 8 = 0$

Solution:

$$\begin{aligned} 4x + 8 &= 0 \\ 4x + 8 - 8 &= 0 - 8 \quad \text{Subtract 8 from both sides} \\ 4x &= -8 \\ \frac{4x}{4} &= \frac{-8}{4} \quad \text{Divide both sides by 4} \\ x &= -2 \end{aligned}$$

Example: Solve $5x - 2 = 3x + 8$

Solution:

$$\begin{aligned} 5x - 2 &= 3x + 8 \\ 5x - 2 + 2 &= 3x + 8 + 2 \quad \text{add 2 to both sides} \\ 5x &= 3x + 10 \\ 5x - 3x &= 3x - 3x + 10 \quad \text{subtract } 3x \text{ from both sides} \\ 2x &= 10 \\ \frac{2x}{2} &= \frac{10}{2} \quad \text{divide both sides by 2} \\ x &= 5 \end{aligned}$$

Example: Solve $\frac{1}{2}x + \frac{2}{3} = \frac{5}{6}$

Solution:

$$\frac{1}{2}x + \frac{2}{3} = \frac{5}{6}$$
$$6 \cdot \frac{1}{2}x + 6 \cdot \frac{2}{3} = 6 \cdot \frac{5}{6} \quad \text{Multiply by 6 (the LCD) to clear the fractions}$$
$$3x + 4 = 5$$
$$3x = 1$$
$$x = \frac{1}{3}$$

2.2 Formulas

What's Important
- Solving for a specific letter.

Literal equations (sometimes called **formulas**) are equations that relate several things. Some familiar formulas:

$$A = lw \quad \text{(area of a rectangle)}$$

$$A = \frac{1}{2}bh \quad \text{(area of a triangle)}$$

$$d = rt \quad \text{(distance is rate times time)}$$

We occasionally need to solve an equation like this for a particular variable. The technique used is not really any different than what we did in the previous section. Just rearrange things so the variable you're interested in ends up by itself on one side. Treat the other variables as though they were numbers.

Example: Solve $A = \frac{1}{2}bh$ for b.

$$A = \frac{1}{2}bh$$
$$2A = bh \quad \text{Multiply both sides by 2}$$
$$b = \frac{2A}{h} \quad \text{Divide both sides by } h$$

2.3 Story Problems

What's Important
• Setting up and solving story problems.

Story problems are unpopular with students, but popular with teachers. The real point of these problems, though, is not some so-called "real-world" application of mathematics, but to see whether you can pick out the important bits of information from all the words in the problem, and then relate those bits to each other in a meaningful way. (This may be the most important and useful skill you can get in a math course.)

Every book has its own list of steps on how to solve a story problem, so why should this book be different? Here's my list:

1. Read the problem and decide what's the important information and what's "local color". If it seems appropriate, draw a picture.

2. Decide on what the problem is asking you to find and pick a variable to represent it. (The last sentence of the problem is a good place to look for the unknown.)

3. Turn the words into symbols. One thing to keep in mind is that the units on both sides of the equation should be the same. So, for example, if you have miles per hour times hours on one side of the equation, you should also have miles per hour times hours on the other side of the equation.

4. Solve the equation.

5. Check your answer. Does it have the right units? Does it have the right sign? Is it bigger or smaller than what you might expect?

2.3. STORY PROBLEMS

And now for some examples.

A distance-rate-time example: The current in a river has a speed of 5 mph. Suppose a person travels in a boat a certain distance downstream in 2 hours and then it takes 4 hours to travel back upstream to the starting point. How fast would the boat go in still water?

Solution: I'll match the solution parts to the steps I gave above.

1. Important information: The speed of the current is 5 mph. The downstream time is 2 hours. The upstream time is 4 hours. The downstream distance and the upstream distance are the same. When you travel with the current, it adds to your speed, while when you travel against the current, it reduces your speed.

2. The problem wants you to find the boat's speed in still water. Let's call this x.

3. The basic formula we will use is $d = rt$. To help us create the equation, let's organize the pieces in a chart:

	Rate	Time	Distance
Downstream	$x + 5$	2	$2(x + 5)$
Upstream	$x - 5$	4	$4(x - 5)$

    And now if we use the fact that the distances are the same, we get our equation:

    $$2(x + 5) = 4(x - 5)$$

4. Solving the equation:

    $$2(x + 5) = 4(x - 5)$$
    $$2x + 10 = 4x - 20$$
    $$-2x = -30$$
    $$x = 15 \text{ mph}$$

5. Notice that the downstream speed was 20 mph, and so in 2 hours, the boat went 40 miles. Meanwhile, the upstream speed was only 10 mph, and so in 4 hours, the boat again went 40 miles.

A business example: The book introduces some business terminology for its example, so I'll do the same.

There are two kinds of costs: *Fixed costs*, which are the same no matter how many items you make or sell (like electricity or rent), and *variable costs*, which change depending on how many items you make or sell (like materials or labor). In general,

$$\text{total cost} = \text{variable cost} + \text{fixed cost}$$

Then *revenue* is the money you get from selling things. Its equation is:

$$\text{revenue} = \text{price} \times \text{quantity}$$

Finally, the *break-even point* is where total cost equals revenue.

The problem: Acme Widgets finds that it costs $5.00 to make each widget. The monthly fixed cost is $1000. If each widget sells for $7.00, how many widgets must be sold in a month to break-even?

Solution: I'll match the solution parts to the steps I gave above.

1. Important information: The fixed cost is $1000. The variable cost is $5.00 per widget. The price per widget is $7.00.

2. The problem wants you to find the number of widgets you need to make and sell in a month in order to break even. Let's let x stand for this quantity.

3. We'll use the facts that the break-even point is where total cost equals revenue, that total cost is fixed cost plus variable cost and that revenue is price times quantity. Then the total cost side will look like: $5x + 1000$, the revenue side will look like $7x$, so our equation will be:
$$7x = 5x + 1000$$

4. Solve the equation:
$$7x = 5x + 1000$$
$$7x - 5x = 5x - 5x + 1000$$
$$2x = 1000$$
$$x = 500 \text{ widgets}$$

5. Check your answer: The cost to make 500 widgets in one month is $5(500) + 1000 = \$3500$ while the revenue from selling 500 widgets is $7(500) = \$3500$. The two numbers are the same, so we've found the break-even point.

2.4 Inequalities

> What's Important
>
> - Solve inequalities and give your answer either as a set or as a number line graph.

A **linear inequality** looks just like a linear equation except that it has a greater than or less than sign between the two sides rather than an equals sign. The same techniques we used to solve linear equations will work in solving linear inequalities, but with one important twist: Adding or subtracting the same number from both sides is okay. Multiplying or dividing by the same *positive* number is okay. The twist is what happens when you multiply or divide by the same *negative* number. If you do this, you must flip the inequality sign around (turn greater than into less than or vice versa). A final note before we look at examples: the solution set to an inequality will usually be a whole range of values, not just a single value as for an equation.

Example: Solve $3x + 4 > 7$

$$3x + 4 > 7$$
$$3x + 4 - 4 > 7 - 4 \quad \text{subtract 4 from both sides}$$
$$3x > 3$$
$$\frac{3x}{3} > \frac{3}{3} \quad \text{divide both sides by 3}$$
$$x > 3$$

Written as a set, our solution would be $\{x | x > 3\}$ and as a graph, the solution would look like

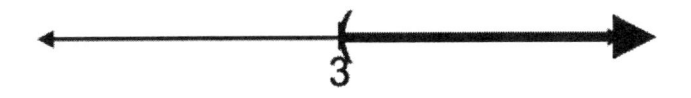

Example: Solve $4 - 3(x+2) \leq 7$

$$4 - 3(x+2) \leq 7$$
$$4 - 3x - 6 \leq 7$$
$$-3x - 2 \leq 7$$
$$-3x \leq 9$$
$$x \geq -3 \quad \text{we divided by } -3, \text{ so we had to flip the sign}$$

Written as a set, the solution is $\{x | x \geq -3\}$ and as a graph:

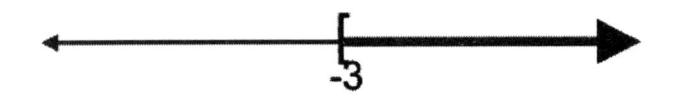

Another kind of linear inequality you'll often run across is called a **compound inequality**. Here's an example:
$$4 < 5 + 2x \leq 7$$
Here's another:
$$5 + 2x < 4 \text{ or } 5 + 2x > 7$$

In both cases, we've combined two simple inequalities together, in the first case using "and" and in the second case using "or".

Compound inequalities are solved the same way as the inequalities we've looked at already. Just do the same operations to all sides.

Example: Solve $4 < 5 + 2x \leq 7$

$$4 < 5 + 2x \leq 7$$
$$4 - 5 < 5 - 5 + 2x \leq 7 - 5$$
$$-1 < 2x \leq 2$$
$$\frac{-1}{2} < \frac{2x}{2} \leq \frac{2}{2}$$
$$-\frac{1}{2} < x \leq 1$$

The solution set is $\{x | -1/2 < x \leq 1\}$ and the graph is:

2.5. ABSOLUTE VALUE EQUATIONS AND INEQUALITIES

Example: Solve: $5 + 2x < 4$ or $5 + 2x > 7$

$$5 + 2x < 4 \quad \text{or} \quad 5 + 2x > 7$$
$$2x < -1 \quad \text{or} \quad 2x > 2$$
$$x < -\frac{1}{2} \quad \text{or} \quad x > 2$$

The solution set is $\{x | x < -1/2 \text{ or } x > 2\}$ and the graph is

2.5 Absolute Value Equations and Inequalities

What's Important
- Solve absolute value equations.
- Solve absolute value inequalities.

The basic form for an absolute value equation is

$$|a| = p$$

where p is a positive number and a is some expression involving a variable.

Example: $|4x + 7| = 5$

Absolute value equations are solved by changing them into two equations without absolute values, like this:

$$|a| = p \text{ turns into } a = p \text{ or } a = -p$$

Example: Solve $|4x + 7| = 5$

Change this equation to

$$4x + 7 = 5 \qquad \text{or} \qquad 4x + 7 = -5$$
$$4x = -2 \qquad\qquad\qquad 4x = -12$$
$$x = -\frac{1}{2} \qquad\qquad\qquad x = -3$$

A somewhat more complicated-looking equation has absolute values on both sides of the equation, but it's not really any harder to solve. Just choose one of the sides to play the role of "p" as above.

Example: Solve $|4x + 7| = |x + 3|$

Change this equation to

$$4x + 7 = x + 3 \qquad \text{or} \qquad 4x + 7 = -(x + 3)$$
$$4x + 7 = x + 3 \qquad\qquad\qquad 4x + 7 = -x - 3$$
$$3x + 7 = 3 \qquad\qquad\qquad 5x + 7 = -3$$
$$3x = -4 \qquad\qquad\qquad 5x = -10$$
$$x = -\frac{4}{3} \qquad\qquad\qquad x = -2$$

Absolute value inequalities come in two basic forms: $|a| < p$ and $|a| > p$. To solve them, you turn them into a compound inequality without the absolute value signs:

$$|a| < p \text{ turns into } -p < a < p$$

$$|a| > p \text{ turns into } a < -p \text{ or } a > p$$

2.5. ABSOLUTE VALUE EQUATIONS AND INEQUALITIES

Example: Solve $|4x - 3| < 5$

Change this inequality into

$$-5 < 4x - 3 < 5$$
$$-2 < 4x < 8$$
$$-\frac{1}{2} < x < 2$$

The solution set is $\{x| -1/2 < x < 2\}$ and the graph is

Example: Solve $|3x - 2| \geq 4$

Change this inequality into

$$3x - 2 \leq -4 \qquad \text{or} \qquad 3x - 2 \geq 4$$
$$3x \leq -2 \qquad\qquad\qquad\qquad 3x \geq 6$$
$$x \leq -\frac{2}{3} \qquad\qquad\qquad\qquad x \geq 2$$

The solution set is $\{x|x \leq -2/3 \text{ or } x \geq 2\}$ and the graph is

2.6 Do-It-Yourself Formulas

> **What's Important**
> - Use a description involving direct, inverse or joint variation to create a formula.

A formula describes how two quantities are related to each other. But there are several different ways this can happen. First, we can have a situation like this:

$$y = 4x$$

In this case, as x increases, so does y, and vice versa. The usual way to describe this is to say that "y is directly proportional to x" or "y varies directly as x".

Second, we could have a situation like this:

$$y = \frac{4}{x}$$

In this case, as x increase, y decreases and vice versa. The usual way to describe this is to say that "y is inversely proportional to x" or "y varies inversely as x".

And then, of course, we can have situations involving more than two variables, such as:

$$y = 4xw$$

or

$$y = \frac{4x}{w}$$

For the first one we would say "y is directly proportional to x and w" or "y varies jointly as x and w", while for the second case we would say something like "y is directly proportional to x and inversely proportional to w" or "y varies directly as x and inversely as w".

Finally, I should say something about that number in the formula. It's called the *constant of proportionality* or *constant of variation* and tells us how the variables are related numerically.

Solving variation problems is really just a matter of translating the words about direct variation, inverse variation and joint variation into an equation with the variables in the right place. The basic rule is that the quantities that vary directly go in the numerator while the quantities that vary inversely go in the denominator. And don't forget to multiply everything by a constant of variation (often denoted by "k").

Example: Suppose y is directly proportional to x and $y = 4$ when $x = 2$. Find y when $x = 7$.

Solution: Solving a problem like this one takes three steps:

2.6. DO-IT-YOURSELF FORMULAS

1. Create the formula.
2. Use the given numbers to find the constant of variation.
3. Answer the question.

Step 1: y is directly proportional to x, so the formula must be $y = kx$.

Step 2: Put 4 in for y and 2 in for x in the formula and solve for k. So $4 = k(2)$, so $k = 2$. So the actual formula is $y = 2x$.

Step 3: Put 7 in for x and solve for y. So $y = (2)(7) = 14$.

Example: Suppose y is inversely proportional to x and $y = 4$ when $x = 2$. Find y when $x = 7$.

Solution: Let's follow the same steps:

Step 1: y is inversely proportional to x, so $y = \dfrac{k}{x}$.

Step 2: Put 4 in for y and 2 in for x and solve for k. So $4 = \dfrac{k}{2}$, so $k = 8$. So the actual formula is $y = \dfrac{8}{x}$.

Step 3: Put 7 in for x and solve for y. So $y = \dfrac{8}{7}$.

Example: Suppose y is directly proportional to x and w and inversely proportional to z. If $y = 4$ when $x = 2$, $w = 1$ and $z = 3$, find y when $x = 1$, $w = 2$ and $z = 3$.

Solution: We'll still go through the same steps:

Step 1: y is directly proportional to x and w, so they'll go in the numerator multiplied together. y is inversely proportional to z, so it will go in the denominator. So the formula looks like $y = k\dfrac{xw}{z}$.

Step 2: Put 4 in for y, 2 in for x, 1 in for w and 3 in for z and solve for k. So $4 = k\dfrac{(2)(1)}{3}$, so $k = 6$. So the actual formula is $y = \dfrac{6xw}{z}$.

Step 3: Put 1 in for x, 2 in for w and 3 in for z and solve for y. So $y = \dfrac{(6)(1)(2)}{3} = 4$.

Chapter 3

The Coordinate Plane and Functions

In this chapter, we'll start to look at two of the most important ideas in algebra. First, how to draw pictures in math (this is usually called **graphing**) and second, rules for matching things up in a special way (**functions**).

3.1 Ordered Pairs

What's Important

- What is an ordered pair?
- Identify domain and range.

As the name suggests, an **ordered pair** is two numbers written in a specific order. For example $(5, 2)$ is an ordered pair. So is $(2, 5)$. But since order matters, these are different ordered pairs. On the other hand, $(5, 4/2)$ is the same ordered pair as $(5, 2)$ since the numbers are the same ($5 = 5$ and $4/2 = 2$) and they're in the same order (5 is first, 2 is second). Finally, ordered pairs are always written using parentheses, so you could say $5, 2$ is a pair of numbers, but not an ordered pair.

Now let's make sets of ordered pairs (we'll call these **relations**). For example, consider the set $\{(5, 2), (2, 5), (3, 1), (-7, 1482)\}$. This is a relation containing four ordered pairs. We'll group all of the first numbers in each pair together into one set (the **domain** of the relation) like so:

$$\text{Domain} = \{5, 2, 3, -7\}$$

We'll also group all of the second numbers in each pair into a set (the **range** of the relation):

$$\text{Range} = \{2, 5, 1, 1482\}$$

And here's another example:

$$\text{Relation} = \{(1,1), (2,4), (3,9), (4,16), (5,25)\}$$
$$\text{Domain} = \{1, 2, 3, 4, 5\}$$
$$\text{Range} = \{1, 4, 9, 16, 25\}$$

3.2 The Rectangular Coordinate System

What's Important

- Plotting points.

- Finding the coordinates of points.

The number line is good for graphing equations and inequalities that have only one variable. Most equations that are used in algebra have at least two variables, so to graph them we need to use two dimensions. The particular system we'll use is called **rectangular coordinates** or **Cartesian coordinates** (in honor of the French philosopher René "I think, therefore I am" Descartes. To set it up, take two number lines and cross them at right angles at their 0 points so that numbers increase left to right and bottom to top.

The plane ends up divided into four sections, called **quadrants**. The quadrants are numbered (using Roman numerals) from I to IV counterclockwise starting from the upper right. The horizontal number line is called the **x-axis** and the vertical number line is called the **y-axis**. The place where they cross is called the **origin**. For a picture of everything I've just said, see the first picture on the next page (Figure 3.1).

The coordinates of points are given by ordered pairs. The first number is called the **x-coordinate** and gives the horizontal position (positive is to the right of the origin, negative is to the left), the second number is the **y-coordinate** and gives the vertical position (positive is above the origin, negative is below). For example, the point $(3, 2)$ (point A on the graph below) would be three places to the right and two places above the origin, while the point $(-2, -1)$ (point B on the second graph below (Figure 3.2)) would be two places to the left and one place below the origin.

3.2. THE RECTANGULAR COORDINATE SYSTEM

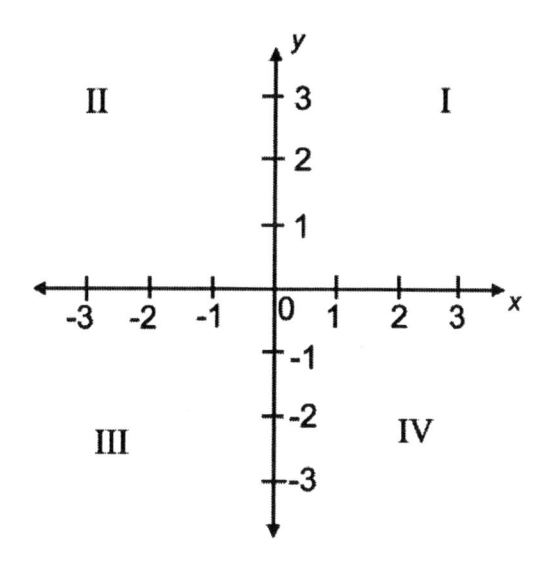

Figure 3.1: Rectangular coordinates

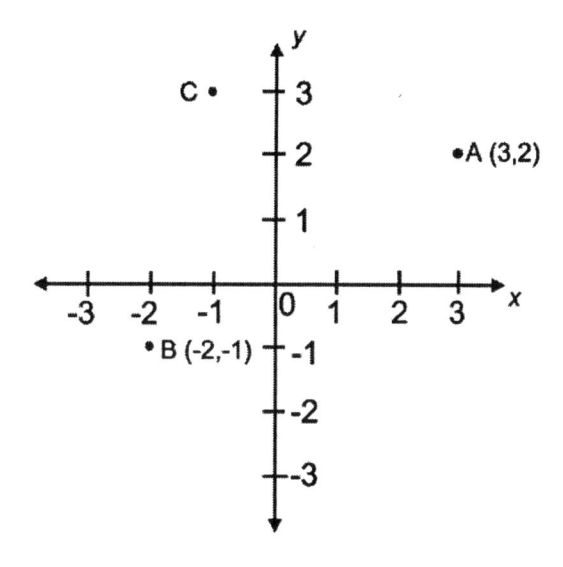

Figure 3.2: Some points

And what would be the coordinates of point C? Well, it's one place to the left and three places above the origin, so its coordinates would be $(-1, 3)$.

3.3 What is a Function?

> **What's Important**
>
> - Decide whether something is a function.
> - Evaluate a function.

Perhaps the most important idea you'll run across in math is the idea of a "function". There are lots of ways to define what a function is, but my favorite way to define it is "a function is a rule for matching something in one set (the **domain**) with *exactly* one thing in a second set (the **range**)". What does this mean in practice? Well, if you give me a rule for matching things up, I can tell you whether that rule describes a function.

Example: The domain is the set of people in the United States, the range is the set of 9-digit numbers. The rule is to match a person with his or her social security number. Is this a function? And the answer is...Yes. A person will only have one social security number, so we've matched something in the domain with exactly one thing in the range.

Example: The domain is the set of car manufacturers, the range is the set of car models. The rule is to match a manufacturer with the model of car it makes. Is this a function? And the answer is...No. In general, a car manufacturer will make several different models of cars, so we can't match something from the domain with only one thing in the range.

There are other ways to represent possible functions. These include:

1. As a set of ordered pairs (like in Section 3.1)
2. As a graph
3. As an expression or formula

Example: Consider the set $\{(5,2),(2,5),(3,1),(-7,1482)\}$. Is this a function? To answer this question, you have to figure out that the rule here is the one that matches the first number in an ordered pair with the second number in that same pair. With that in mind, the answer is...Yes. If you check all of the pairs, any particular first number gets matched with only one second number.

Example: Consider the set $\{(5,2),(2,5),(5,1),(-7,1482)\}$. Is this a function? And the answer is...No. If you check the ordered pairs here, you'll see that in the first pair 5 gets matched with 2 while in the third pair, 5 gets matched with 1. So we have a particular first number that gets matched with more than one second number, so it can't be a function.

3.3. WHAT IS A FUNCTION?

Example: Consider this picture:

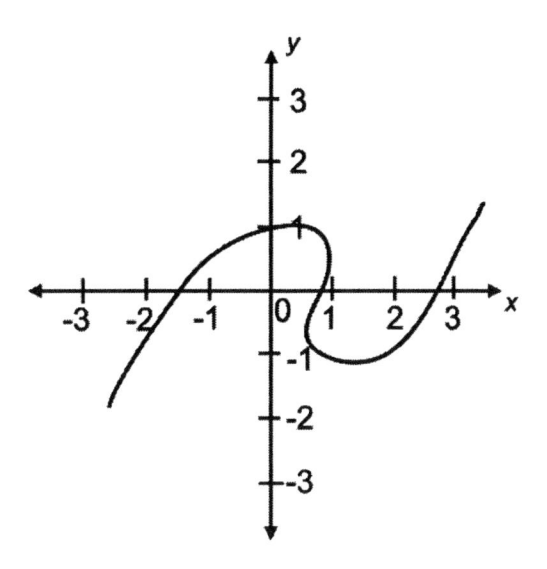

Is this a function? Again, we have to first figure out what the rule is. Since a graph is really just a collection of ordered pairs, the rule is again matching the first number (the x–coordinate) with the second number (the y–coordinate) for each point. Then how do we tell whether a particular x–coordinate gets matched up with only one or more than one y–coordinate? A visual way to do this is to try drawing vertical lines. If you can draw a vertical line that intersects the graph more than once, then it's not a function; otherwise, it is a function. (This is called the **vertical line test**.) So for the graph above, it turns out a vertical line drawn just to the left of the 1 on the x-axis will cross the graph three times, so it can't be a function.

For students who have had previous math classes, using an expression or formula to represent a function is the most familiar method of them all. For example, $f(x) = x^2 - 4x + 7$ is a function. You can think of the "$f(x)$" part as being the name of the function. (You can call your functions something else if you want. How about a function called "Aloysius"?) The "x" in parentheses tells you what the variable is in the function. The values that go in for x are from the domain, while the values that come out of the function are in its range.

Functions given by expressions are the easiest ones to evaluate. Just take the number in the parentheses, put it in for x everywhere and calculate.

Example: Let $f(x) = x^2 - 4x + 7$. Then $f(2) = 2^2 - 4(2) + 7 = 3$ and $f(-2) = (-2)^2 - 4(-2) + 7 = 19$.

3.4 Reading a Graph

> **What's Important**
>
> - Use a graph to evaluate a function.
> - Use a graph to find the intercepts of a function.

Since drawing a graph can be a useful and compact way to present information ("a picture is worth a thousand words" and all that), you need to be able to read the graph to get the information you want. For the moment, we'll just look at how to get function values. Later on, we'll look for more things.

So let's look at this graph:

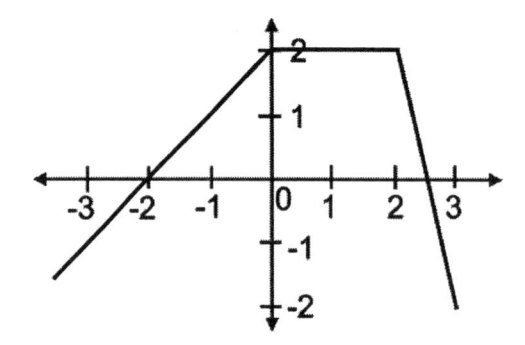

This is the graph of some function (you can check that it passes the vertical line test) which we'll call $f(x)$. The question we want to ask is: What is $f(1)$? To answer that question, find the point on the graph which has x–coordinate equal to 1 and figure out the y-coordinate. The y–coordinate will be the function value that we're looking for. That point is $(1, 2)$, so we can say $f(1) = 2$. So try another one: What is $f(-1)$? The point on the graph that has x-coordinate equal to -1 is $(-1, -1)$, so $f(-1) = -1$. Finally, what is $f(-2)$? Here the point we want is $(-2, 0)$, so $f(-2) = 0$.

The last point we found is a very special point where the graph crosses the x–axis. It's called an **x–intercept**, and we'll often want to find it. Notice that this graph has a second x–intercept at $(2.5, 0)$. A related point that we're also interested in is where the graph crosses the y-axis, called, naturally, the **y–intercept**, and on this graph it's the point $(0, 2)$. An important thing to keep in mind about functions and intercepts is that a function can have lots of x–intercepts (or maybe none at all), but it will never have more than one y–intercept. (Why?)

By the way, we can read the graph "backwards" if we want, meaning, we can start with a value for the function and ask what x values we would put into the function. (There may be several answers.) For example, for the graph above, for what values of x does $f(x) = -1$? To answer this question, find the points on the graph where the y–coordinate is -1 and figure out the x–coordinates of those points. For this graph, the points are $(-3, -1)$ and (approximately) $(2.8, -1)$. So the answer to our question is $x = -3$ and $x = 2.8$.

Chapter 4

All About Lines

In this chapter, we'll look at lines in various ways. We'll start with how to graph a line, then how to find an equation of a line, and then finish up with a look at how the graph of a line can help us solve some kinds of problems.

4.1 Drawing Lines

What's Important

- Find the intercepts of a line.
- Graph a line.

Back in Chapter 2 we looked at things like this: $4x - 4 = 12$. We called this a "linear equation in one variable". Now we'll start to look at things like this: $4x - 2y = 8$. This is a **linear equation in two variables**. Recall that the solution to the first type of equation was a single number. For the equation above, the solution would be $x = 4$. For the second type of equation, the solution will be a whole set of ordered pairs. For example, one possible solution is the ordered pair $(1, -2)$. So how do we get a solution? The easiest way is to pick any number, substitute it in to the equation for x and then solve for y. So if I put 1 in for x I get:

$$4(1) - 2y = 8$$
$$4 - 2y = 8$$
$$-2y = 8 - 4$$
$$-2y = 4$$
$$y = -2$$

Since the solution is going to be a set of ordered pairs, we could plot the ordered pairs as points and get a graph. What do you think the graph will look like? (Hint: It's a *linear* equation.) Yes, the graph is going to be a line. If we want to graph our line $4x - 2y = 8$ we actually need two different points, so we need another solution. One possibility would be $(3, 2)$. Then the graph will look like this (we put arrowheads on the ends to show it keeps on going):

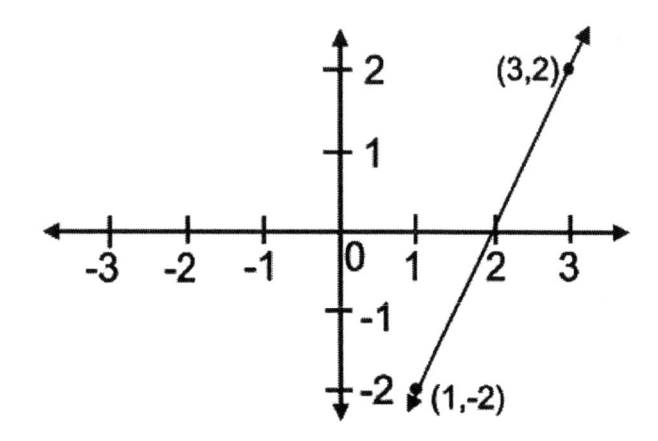

So to graph a line, find any two points on the line (find any two solutions of its equation), plot them and connect the dots.

Two nice points to find are the intercepts of the line. In Section 3.4 we looked at how to find the intercepts by looking at the graph. Now we want to do the opposite: find the intercepts so that we can draw the graph. Here's how:

- To find the x–intercept, put zero in for y and solve for x.

- To find the y–intercept, put zero in for x and solve for y.

4.1. DRAWING LINES

Example: Find the intercepts and sketch the graph of $3x + y = 3$.

x–intercept:

$$3x + 0 = 3$$
$$3x = 3$$
$$x = 1$$

y–intercept:

$$3(0) + y = 3$$
$$y = 3$$

So the x–intercept is $(1, 0)$ and the y–intercept is $(0, 3)$ and the graph looks like

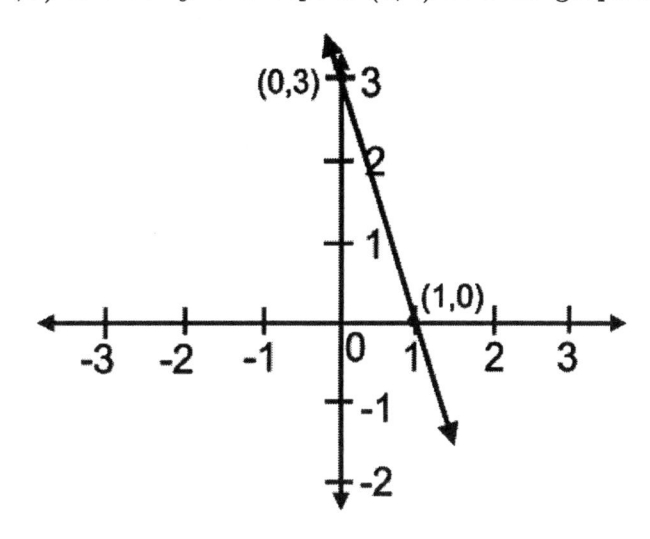

Some lines only have one intercept. For example, try the line $4x + 3y = 0$. If you find the intercepts as we did in the last example, you'll find that both the x–intercept and the y–intercept are $(0, 0)$. So if you want to graph this line, you'll need to find some other point that's not an intercept. So pick another number to put in for x (for example, 3) and solve for y (to get, for example $y = -4$).

Some other kinds of lines that have only one intercept are horizontal and vertical lines. You can recognize these lines by their special equations. A horizontal line can always be written in the form $y =$ number and a vertical line can always be written in the form $x =$ number. To graph them, draw a horizontal or vertical line through that number on the y– or x–axis respectively.

Example: The graph of $x = 2$ looks like:

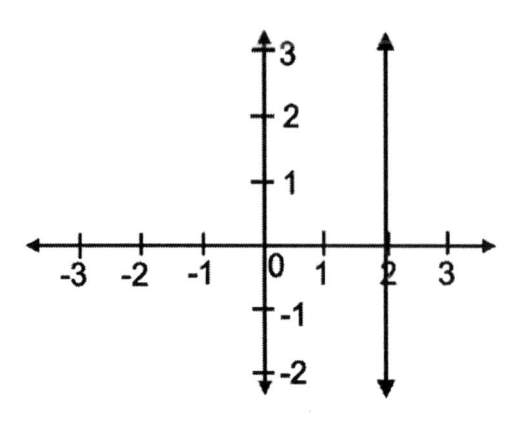

4.2 The Slope of a Line

> **What's Important**
> - Given two points, find the slope of a line.
> - Decide if two lines are parallel, perpendicular, or neither.

After you've drawn enough lines, you notice there are four basic ways lines can behave:

1. The line can slant up from left–to–right.

2. The line can slant down from left–to–right.

3. The line can be horizontal.

4. The line can be vertical.

It turns out that by calculating one number, you can figure out which of these ways your line will behave, and what's more, if the line slants, you can get an idea of how fast it goes up or down. This one number is called the **slope** of the line. It's usually denoted by a lowercase m and it's defined as the vertical distance between two points (the **rise** or **change in y** divided by the horizontal distance between those same two points (the **run** or **change in x**). In other words, if we have two points on the line (x_1, y_1) and (x_2, y_2), then the slope will be:

$$\text{slope} = m = \frac{\text{rise}}{\text{run}} = \frac{\text{change in } y}{\text{change in } x} = \frac{y_2 - y_1}{x_2 - x_1}$$

4.2. THE SLOPE OF A LINE

Example: The slope of the line through the points $(4, 2)$ and $(7, -1)$ is

$$m = \frac{-1-2}{7-4} = \frac{-3}{3} = -1$$

Note that it actually doesn't matter which point we choose as the first point and which as the second as long as we're consistent. So we could have done the last example like this:

$$m = \frac{2-(-1)}{4-7} = \frac{2+1}{4-7} = \frac{3}{-3} = -1$$

Example: The slope of the line through the points $(4, 2)$ and $(4, 7)$ is

$$m = \frac{7-2}{4-4} = \frac{5}{0} = \text{undefined}$$

So how does slope tell us how the line behaves? Like so:

- If the slope is positive, the line slants up from left–to–right. The larger the slope, the faster it goes up.

- If the slope is negative, the line slants down from left–to–right. The more negative the slope, the faster it goes down.

- If the slope is zero, the line is horizontal.

- If the slope is undefined, the line is vertical.

Slope can also be used to figure out how two lines are related to each other. In particular, it can tell us whether two lines are **parallel** (never intersect) or **perpendicular** (intersect at right angles).

- Two lines are parallel if and only if they have the same slope. (Symbolically, we would say $m_1 = m_2$.)

- Two lines are perpendicular if and only if their slopes are negative reciprocals of each other. (Symbolically, we would say $m_2 = -1/m_1$ or $m_1 \cdot m_2 = -1$.) Horizontal lines and vertical lines are also perpendicular to each other.

Example: Is the line through $(4,7)$ and $(5,2)$ parallel, perpendicular or neither to the line through $(0,0)$ and $(5,1)$?

Solution: Find the slopes of the two lines and compare them. The slope of the first line is $m_1 = \frac{2-7}{5-4} = \frac{-5}{1} = -5$ while the slope of the second line is $m_2 = \frac{1-0}{5-0} = \frac{1}{5}$. These two slopes are negative reciprocals of each other, so the lines must be perpendicular.

One final use for slope is that it can help us to graph a line quickly. The method is to start with one point, then use the fact that slope is change in y over change in x to find a second point. If the slope is positive go up and to the right, if it's negative, go down and to the right. (What if the slope is just an integer, not a fraction? Then the change in x is just 1.)

Example: The graph of the line through $(-1,-1)$ with slope 3 can be drawn by plotting $(-1,-1)$, then going 3 places up and one place to the right for the second point. So the graph looks like:

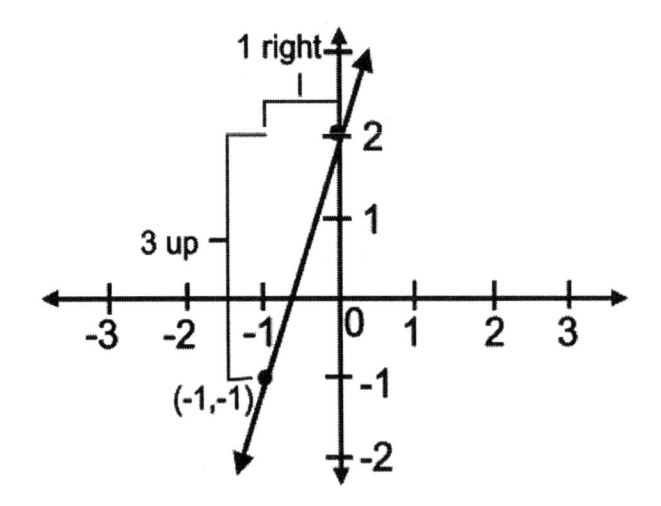

4.3 Equations of Lines

> **What's Important**
> - Write the equation of a line in point–slope form.
> - Write the equation of a line in slope–intercept form.
> - Find equations of parallel and perpendicular lines.

There are four standard ways to write an equation of a line:

1. *General Form*: $Ax + By + C = 0$. This form is popular with mathematicians.

2. *Standard Form*: $Ax + By = C$. This form is popular with textbook writers and math teachers (especially for tests).

3. *Point–Slope Form*: $y - y_1 = m(x - x_1)$, where m is the slope and (x_1, y_1) is a point on the line. This form is the easiest one to use to just write down the equation of the line.

4. *Slope-Intercept Form*: $y = mx + b$, where m is the slope and $(0, b)$ is the y-intercept. This form is handy when you need to figure out the slope of some line or if you want to graph the line.

All of these forms can be turned into one another just by rearranging things or by solving for y. So with that in mind, I'll usually give the answers to the examples in point–slope form (less work that way).

To actually find an equation of a line you always need to know two things: the slope of the line (we looked at slope in the previous section) and a point on the line. (Notice that if we have these, we can put them in the point–slope form right away and we're done.) So in most problems of this type, it all comes down to deciding on how to figure out what the slope is and what point to use.

Example: Find an equation of the line through the point $(4, 2)$ and $(5, -7)$.

Solution: First, find the slope: $m = \dfrac{-7 - 2}{5 - 4} = \dfrac{-9}{1} = -9$. Then pick one of the points you were given and put the numbers in the point–slope form. If we use the first point we get: $y - 2 = -9(x - 4)$ while if we use the second point we get $y - (-7) = -9(x - 5)$. Note that though the equations are different, they actually are the same line. The easiest way to see this is to put them in slope–intercept form by solving for y:

$$y - 2 = -9(x - 4)$$
$$y - 2 = -9x + 36$$
$$y = -9x + 38$$

$$y - (-7) = -9(x - 5)$$
$$y + 7 = -9x + 45$$
$$y = -9x + 38$$

Example: Find an equation of the line through the point $(4, 2)$ that is parallel to the line $4x + 3y = 7$.

Solution: For a problem like this, we need to use the slope of the given line to figure out the slope of the line we're trying to find. To find the slope of the given line, put it in slope–intercept form (solve for y). It becomes $y = (-4/3)x + (7/3)$. In this form, the number in front of the x is the slope, so the slope is $-4/3$. Now recall from Section 4.2 that parallel lines have the same slope, so the slope of the line we're trying to find must also be $-4/3$. And now we put that value of the slope and the point $(4, 2)$ into the point–slope form and we have our equation: $y - 2 = (-4/3)(x - 4)$.

I said that the slope–intercept form is useful if we want to graph the line. The idea is that if we have a line in this form, then it not only tells us the slope, but also the y–intercept, which we then can use as a starting point in our graph. For example, the slope of the line $y = (-4/3)x + 3$ is $-4/3$ and the y–intercept is $(0, 3)$. So to graph $y = (-4/3)x + 3$ we would plot the point $(0, 3)$, then go 4 places down and 3 places to the right to get our second point.

The slope–intercept form has one other use: A **linear function** has the form $f(x) = mx + b$, which is exactly the slope–intercept form. So questions about linear functions are really just questions about lines in slope–intercept form.

Example: What is the slope of the function $f(x) = 3x - 2$?

Solution: Since the function is in slope–intercept form, the number in front of the x is the slope. So the slope is 3.

4.4 Solving Things with Graphs

What's Important

- Use a graph to solve a linear equation.
- Use a graph to solve a linear inequality.

Using graphs to solve linear equations or inequalities is really just graphing lines. (Linear inequalities will involve a little more, however.) So for example, if we want to use a graph to solve the equation $3x + 1 = -2$, we first treat each side as the equation of a line:

Left Side   Right Side
$y = 3x + 1$ $y = -2$

Then we plot both of these lines and see where they intersect:

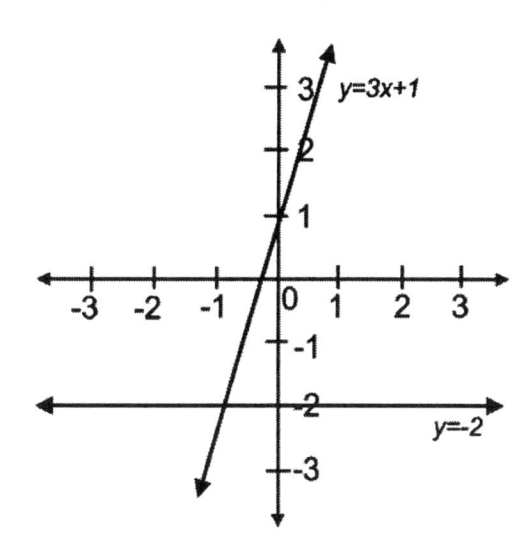

In this case, the two lines intersect at the point $(-1, -2)$, so the solution would be $x = -1$.

At this point, I'll throw in my opinion: Unless you are absolutely required to, always solve your linear equations algebraically rather than graphically. You'll have fewer problems that way. But I should also mention that you'll see this idea of graphing two lines and seeing where they intersect reappear in Section 5.1.

Solving a linear inequality graphically also starts by treating the two sides as two separate lines and graphing them. For example, if we want to solve $3x + 1 > -2x - 2$ then we would start by graphing the lines $y = 3x + 1$ and $y = -2x - 2$. Then we draw a vertical line through their intersection point (dotted in this case because it's a strict inequality, but solid if we have ≥ or ≤) and shade the x–axis on the side of the vertical line where the line $y = 3x + 1$ is above the line $y = -2x - 2$. The final picture looks something like this:

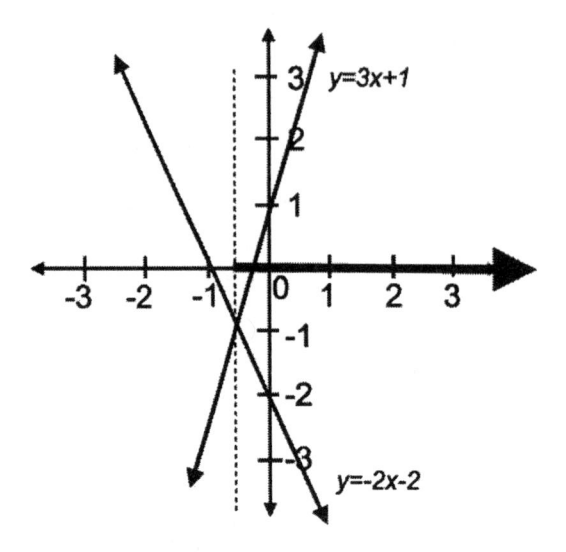

Again, in my opinion, you're better off solving linear inequalities algebraically (though see Section 5.4 for inequalities that you must solve graphically).

4.5 More Things to Solve with Graphs

What's Important

- Use a graph to solve an absolute value equation.
- Use a graph to solve an absolute value inequality.

Continuing with the theme of the previous section, we'll use graphs to solve equations and inequalities. Again we'll be doing this by graphing two things, but in this case one of the things will be an absolute value function. So first we need to see how to graph $f(x) = |x|$. Start by making a table of values:

x	-2	-1	0	1	2
$f(x)$	2	1	0	1	2

Then plot the points and sketch the graph:

4.5. MORE THINGS TO SOLVE WITH GRAPHS

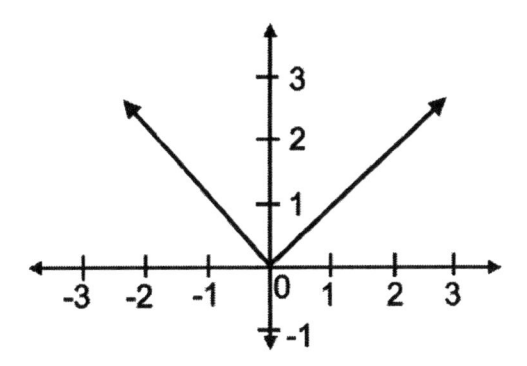

We actually need to be able to graph slightly more complicated absolute value functions, namely, functions that have the form $f(x) = |x - a|$ or $f(x) = |x + a|$. It turns out the graphs have the same shape as $f(x) = |x|$ but they've been moved to the left a spaces (for $f(x) = |x+a|$) or moved to the right a spaces (for $f(x) = |x - a|$). For example, the graph of $f(x) = |x - 2|$ looks like:

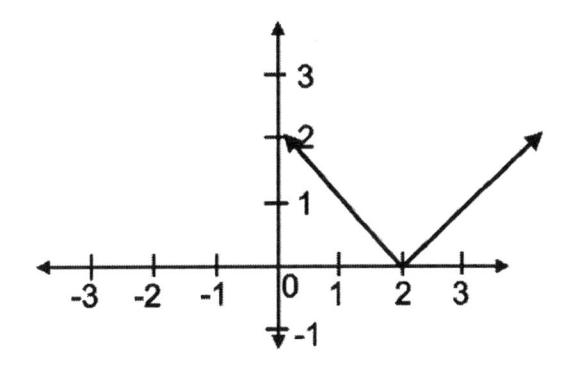

And now for the actual equations and inequalities.

Example: Solve $|x - 2| = 1$
Solution: Graph $f(x) = |x - 2|$ and $y = 1$ and see where they intersect:

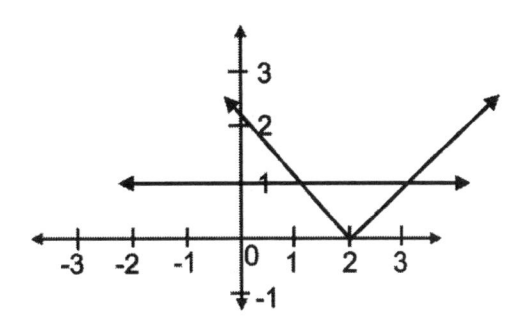

They intersect in two places, at the points $(3,1)$ and $(1,1)$, so the answers are $x = 1$ and $x = 3$.

Example: Solve $|x - 2| > 1$
Solution: Again, we graph $f(x) = |x - 2|$ and $y = 1$ and see where they intersect. Now draw dotted vertical lines through the intersection points and shade the part of the x-axis where the graph of $f(x) = |x - 2|$ is above the graph of $y = 1$. Notice that the answer is in two pieces.

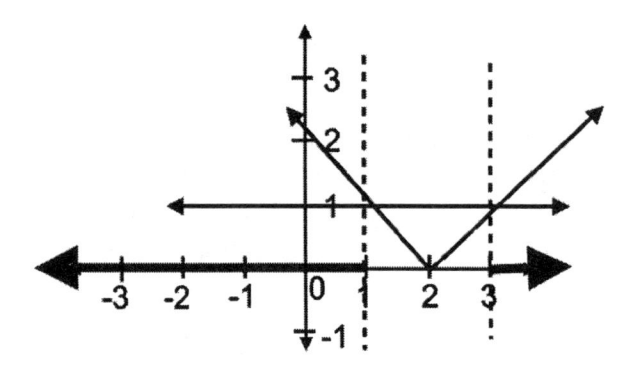

As with linear equations and inequalities, I think you're better off solving them algebraically than graphically.

4.6 Using Graphs

What's Important

- Use a graph to represent "real-world" data.
- Pick out the dependent variable and the independent variable in an equation.

Graphs show up in newspapers, magazines, business presentations, etc., all of the time. They're made the same way that we've made graphs so far: you plot your points and connect the dots (a **line graph**) or don't connect the dots (a **scatter plot**). Perhaps the one part of these graphs that's different from the graphs we've drawn so far is how the axes are set up. First, in many cases, you don't use x and y; instead you may use time and money or something else. Second, we've had nice regular scales so far (each mark represented one unit). For these "real-world" graphs, the marks may represent different amounts of units. For example, if the horizontal axis is time and the vertical money, each mark on the horizontal axis may represent one year while each mark on the vertical axis may represent a million dollars.

4.6. USING GRAPHS

Example: If Acme Widgets makes 100 widgets, it will earn $4000, while if it makes 250 widgets it will make $7000. Find a linear equation that relates the number of widgets w to the revenue R and graph it.

Solution: The words "linear equation" should give you a hint that this problem is really about finding and graphing a line. All that stuff about number of widgets and earnings is really there to give you two points on the line, namely, $(100, 4000)$ and $(250, 7000)$. So find the slope:

$$m = \frac{7000 - 4000}{250 - 100} = \frac{3000}{150} = 20$$

and write down your equation: $R - 4000 = 20(w - 100)$, so $R = 20w + 2000$.

And now to graph it, we have to figure out how to set up the axes. For an equation like this, we call w the **independent variable** because we can choose any value we would like for w. (Well, almost any. For this problem, we probably want $w \geq 0$.) On the other hand, we call R the **dependent variable** because its value *depends* on the value of w. In general, if you've solved an equation for a particular letter, that will be the dependent variable and the other letter will be the independent variable.

What does this have to with graphing? In graphing an equation, you always put the values for the independent variable on the horizontal axis and the values for the dependent variable on the vertical axis.

The other thing to figure out is the scale. How much should each mark stand for? Probably a good scale for this problem is to use units of 100 (widgets) on the horizontal axis and units of 1000 (dollars) on the vertical axis. So our graph will look like this:

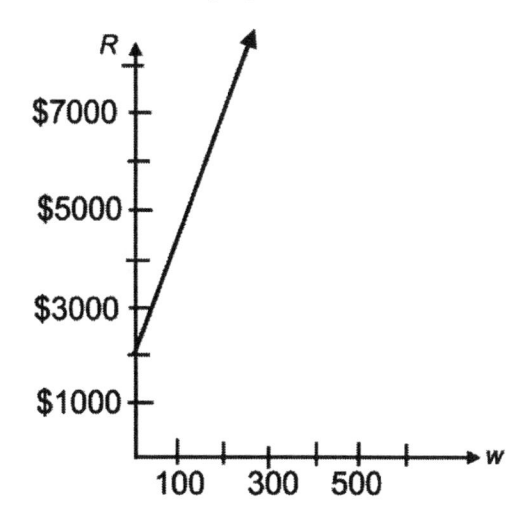

If we had chosen a different scale, our line would look different. For example, if we had made each mark on the w axis worth 50 widgets, the line would not have been so steep.

If we just have two sets of related numbers that we want to graph, it's often good enough just to plot the ordered pairs but not connect them up. As I mentioned above, this is called a **scatter plot**. Scientists often create scatter plots of their experimental data, but they don't stop there. They look at the picture and say "Hey, that looks like a line", or "Hey, that looks like some other curve". (This is called **curve fitting**.) We'll just make a scatter plot and ignore the curve fitting.

Example: A certain unnamed math professor kept track of how his students did on the final exam (worth 200 points) as compared to their final score (900 points possible). Some of the data looks like:

Final Exam	150	130	105	170	125	143
Final Score	700	650	500	820	600	650

A scatter plot for this data looks like the graph below. Do you see any connection between the final exam score and the final class score?

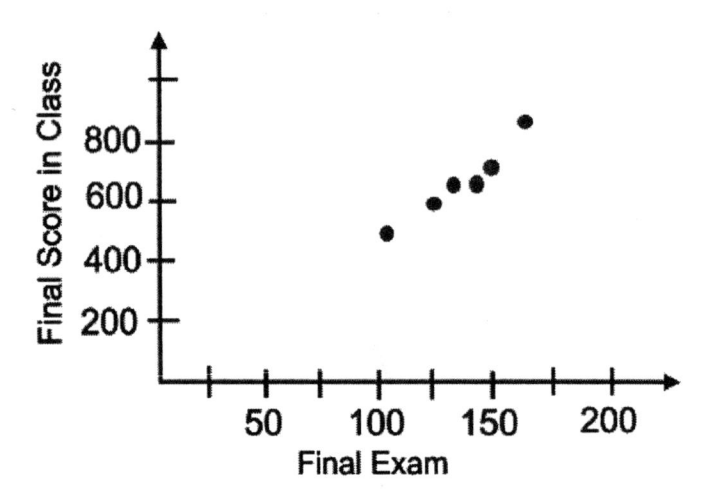

Chapter 5

More Than One Equation

In Chapter 4 we looked at lines, though we only dealt with one line at a time. In this chapter we'll look at more than one equation at the same time, what's usually called a **system of linear equations**.

5.1 Two Lines at the Same Time

What's Important

- Solve a system of two linear equations.

- Describe linear systems as consistent, inconsistent or dependent.

Suppose we have two lines:

$$x + 2y = 4$$
$$2x - y = 3$$

A natural question to ask might be: Are there values for x and y that will work in both equations at the same time? For this pair of lines, the answer is yes. If you try $x = 2$ and $y = 1$, you'll see that these numbers work in both equations. But if you try any other pair of numbers, they may work in one equation or the other, but they won't work in both at the same time.

We spent most of Chapter 4 graphing lines, so let's graph these lines. (See Figure 5.1 at the top of the next page.)

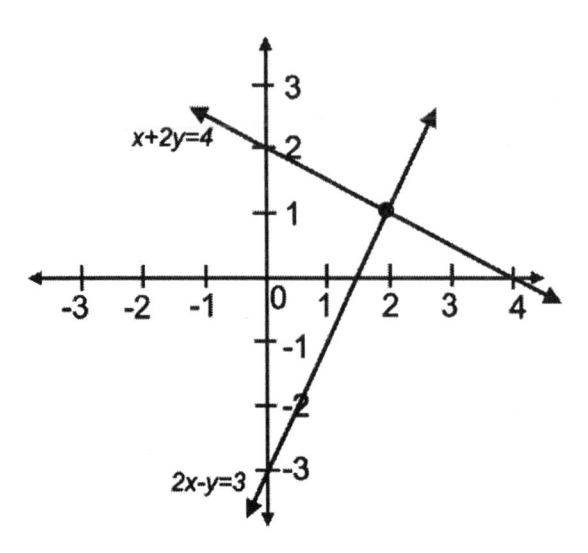

Figure 5.1: $x + 2y = 4$ and $2x - y = 3$

Where do the two lines cross? At the point $(2, 1)$! It turns out that solving a system of equations like this is really about figuring out where two lines cross. So one possible way to solve systems of equations is to graph them and read off the intersection points. However, it's not a terribly good way to do it, since getting the right answer may depend on how well you can draw your graphs and how accurately you can read the coordinates of the point. (How good would your answer be if the intersection point fell between the marks?) However, it's a great way to show the three possible ways two lines can interact:

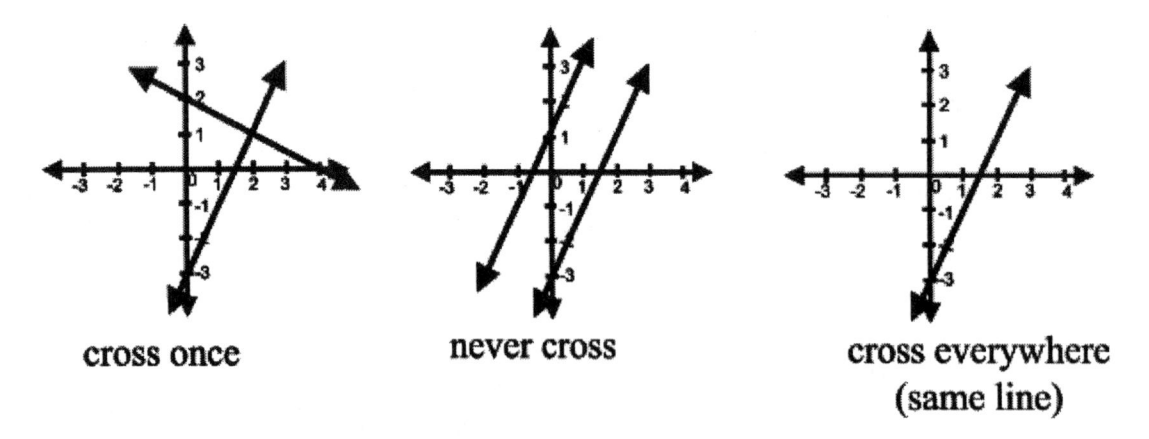

cross once   **never cross**   **cross everywhere (same line)**

These three ways correspond to three possible numbers of solutions to a system of equations:

1. One solution – called a **consistent** system.

5.1. TWO LINES AT THE SAME TIME

2. No solution – called an **inconsistent** system.

3. Infinitely many solutions – called a **dependent** system.

The two other ways to solve a system are both algebraic. The first is called **substitution**, the second is called **elimination**, or sometimes, **elimination by addition**. We'll start with substitution.

1. Take one of your equations and solve it for one of the letters.

2. Take the expression you just found and substitute it in to the other equation.

3. You'll now have an equation that has only one variable, so solve the equation.

4. Finally, take the number you found and put it back into the result of Step 1 to get the value for the other variable.

An example will probably make this clearer:

Example: Solve $\begin{array}{l} x + 2y = 4 \\ 2x - y = 3 \end{array}$

Solution: Solve the first equation for x and you get $x = 4 - 2y$. Now take the $4 - 2y$ and put it in for the x in the second equation:

$$2(4 - 2y) - y = 3$$

Solve this equation: $2(4 - 2y) - y = 3$, so $8 - 4y - y = 3$, so $8 - 5y = 3$, so $-5y = -5$, so $y = 1$. Finally, take put 1 in for y in the expression $x = 4 - 2y$ to get $x = 4 - 2(1) = 4 - 2 = 2$. So the solution is $x = 2$, $y = 1$.

The elimination method will probably sound more complicated, but it isn't really.

1. Pick one of the letters.

2. Multiply one or both of the equations by some number so that the coefficient of the letter you picked is the same in both equations, but has the opposite sign in one of them. (Be sure to multiply *everything* by the number.)

3. Add the equations. (Be sure to add *everything*.)

4. The result has only one variable, so solve the result.

5. Take the value you found in Step 4, put it in one of your original equations and solve for the other letter.

Let's try solving the same system using elimination:

Example: Solve $\begin{array}{c} x + 2y = 4 \\ 2x - y = 3 \end{array}$

Solution: Let's eliminate y. If I multiply the second equation by 2, the first equation will have a 2 in front of the y while the second equation will have a -2 in front of the y:

$$x + 2y = 4$$
$$4x - 2y = 6$$

Now we add the equations and we get $5x = 10$. Solve this and we get $x = 2$. Now take the 2 and put in in for x in the first equation: $2 + 2y = 4$, so $2y = 2$, so $y = 1$. So again, our solution is $x = 2$, $y = 1$.

Let's try a more complicated example:

Example: Solve $\begin{array}{c} 2x + 5y = -3 \\ 3x - 2y = 5 \end{array}$

Solution: This time, let's eliminate the x. To do this, we need to multiply the first equation by 3 and the second equation by -2:

$$6x + 15y = -9$$
$$-6x + 4y = -10$$

Add these together and we get $19y = -19$, so $y = -1$. Now put -1 in for y in the first equation: $2x + 5(-1) = -3$, so $2x - 5 = -3$, so $2x = 2$, so $x = 1$. So the solution is $x = 1$, $y = -1$.

5.2 Story Problems and Systems of Equations

> What's Important
> - Solve story problems by using a system of two equations.

Of course we can use systems of equations to solve story problems. Problems that involve finding two (or more) different quantities are particularly suitable. The classic type of story problem that we use a system of equations to solve is called a *mixing problem*. It's called this because the standard example involves mixing together two chemical solutions to get a new solution (see below for my example). But mixing problems can also appear in other situations. The things to look for:

1. Two or more items.

2. The items are combined together in some way to get a whole.

In solving story problems, the most important part (and usually the hardest part) is setting up the equation. Mixing problems are no different, and, in fact, may seem harder because you now need two (or more) equations. (Side note: I keep saying two or more. The reason for this is that I'm getting you ready for the next section, when you'll have three equations. It will turn out that you'll always want as many equations as you have different quantities to find. So if you're looking for two things, you'll need two equations, three things, three equations, and so on. For the moment, we'll just need two equations.) However, if you read the problem carefully, you'll see that you're always given two different kinds of information, such as total amount and percentages or total number and total revenue. You use each of these to create an equation. Keep this idea in mind: **Have like quantities all the way through the equation!** By this I mean that if you have, for instance, "percentage times amount" as one term in your equation, then all of the terms in your equation should look like "percentage times amount". And now for some examples.

Example: A chemist mixes a 15% salt solution and a 20% salt solution to get 5 liters of an 18% salt solution. How many liters of each of the two solutions were used?

Solution: First, let's notice the two kinds of information mentioned in the problem. First, there's total amount of the final solution: 5 liters. Second, there's the concentration (percentage of salt) of the final solution: 18%. This suggests that for our two equations, one equation will involve amounts of solutions (in liters) and the other will involve concentrations (percentage times amount). So let's pick some variables and set up the equations. We'll let x represent the amount in liters of the 15% salt solution and we'll let y represent the amount in liters of the

20% salt solution. Then if we add the two amounts together, we get the amount of the final solution. So our first equation looks like this:

$$x + y = 5$$

For the second equation, I said above we should use "percentage times amount". (We're actually figuring out the amount of salt in each solution.) The second equation ends up like this:

$$0.15x + 0.20y = 0.18(5)$$

(Remember to change your percentages to decimals!) Notice that all of the terms here look like "percentage times the corresponding amount".

So we now have a system of two equations:

$$x + y = 5$$
$$0.15x + 0.20y = 0.9$$

To solve this system, you can use any method from Section 5.1 that you want, though substitution may be easiest.

Solve the first equation for x:

$$x = 5 - y$$

Substitute in the second equation:

$$0.15(5 - y) + 0.20y = 0.9$$

Multiply out, combine like terms and solve for y:

$$0.75 - 0.15y + 0.20y = 0.9$$
$$0.75 + 0.05y = 0.9$$
$$0.05y = 0.15$$
$$y = 3$$

Finally, solve for x:

$$x = 5 - 3$$
$$x = 2$$

5.2. STORY PROBLEMS AND SYSTEMS OF EQUATIONS

So the answer is 2 liters of the 15% solution and 3 liters of the 20% solution.

Our next example talks about tickets to a concert, but it's still really a mixing problem. Notice that we take two kinds of tickets and combine them together to get the whole audience. That fits the description of a mixing problem that I gave above.

Example: 700 tickets were sold for a concert for a total revenue of $4600. If adult tickets cost $8 and children's tickets cost $5.50, how many of each type of ticket were sold?

Solution: Again, notice the two types of information we're given: the total number of tickets (700) and the total revenue ($4600). So one equation should be about the number of tickets and the other about revenue. So let's let x be the number of adult tickets and y be the number of children's tickets. Add these together and you'll get the total number of tickets, so we have the first equation:

$$x + y = 700$$

For the second equation, remember that we calculate revenue by doing price times quantity. Also remember that we want all of our terms to be like quantities. So our second equation must be:

$$8x + 5.50y = 4600$$

(I know, the terms don't all look the same, but each term is about revenue (money). The first term ($8x$) gives the revenue from adult tickets, the second term ($5.50y$) gives the revenue from children's tickets and the term on the right side (4600) is the total revenue.) So here's our system:

$$x + y = 700$$
$$8x + 5.50y = 4600$$

To solve this system, we'll use substitution again. So solve the first equation for x:

$$x = 700 - y$$

Substitute this in the second equation:

$$8(700 - y) + 5.50y = 4600$$

Multiply out, combine like terms, and solve for y:

$$5600 - 8y + 5.50y = 4600$$
$$-2.50y = -1000$$
$$y = 400$$

Finally, solve for x:

$$x = 700 - 400$$
$$x = 300$$

So 300 adult tickets were sold, and 400 children's tickets were sold.

5.3 Three Equations in Three Variables

> **What's Important**
> - Solve systems of three linear equations by elimination.

Mathematicians always like to generalize things. One popular way to generalize is to try to do something using more variables. So that's what we'll do now, we'll generalize what we've been doing in the last two sections. In other words, we'll solve systems of three linear equations in three variables. (If you want, you can keep going to four equations with four variables and so on, but you're on your own. It's not any harder really than what we'll do, though, just longer.)

The standard way to solve systems of three equations is elimination. It is possible to solve them using substitution, but that gets horribly messy very quickly. Theoretically, you could solve them by graphing, but you have to be able to draw in three dimensions. (The graph of a linear equation with three variables is not a line but a plane.) So here's a couple of examples:

Example: Solve the system

$$x + y + z = 4$$
$$3x - 2y + 2z = 5$$
$$x - y + z = 2$$

Solution: To solve a system like this using elimination, start by picking a letter to eliminate. For example, let's climinate the y. We actually have to eliminate the y from two equations, so we're going to work with two different pairs of equations. In this case, we'll work with the first and second equations and the first and third equations.

5.3. THREE EQUATIONS IN THREE VARIABLES

First and second equations: To eliminate the y, multiply the first equation by 2 and then add them together.

$$\begin{array}{rl} 2(x+y+z=4) & \implies \quad 2x+2y+2z=8 \\ 3x-2y+2z=5 & \quad \underline{3x-2y+2z=5} \\ & \quad 5x+4z=13 \end{array}$$

First and third equations: To eliminate the y, just add the equations together.

$$\begin{array}{r} x+y+z=4 \\ \underline{x-y+z=2} \\ 2x+2z=6 \end{array}$$

Now we're down to this system:

$$\begin{array}{r} x+y+z=4 \\ 5x+4z=13 \\ 2x+2z=6 \end{array}$$

The next step is to ignore the first equation and just consider the second and third equations, which form a system of two equations with two variables, just like we've solved before. We'll continue by eliminating the z. To do this, multiply the third equation by -2 and add it to the second equation:

$$\begin{array}{rl} 5x+4z=13 & \implies \quad 5x+4z=13 \\ -2(2x+2z=6) & \quad \underline{-4x-4z=-12} \\ & \quad x=1 \end{array}$$

And now we work our way back through the equations to find the values for the other variables. Put 1 in for x in the equation $5x+4z=13$ to get z (you can also use the third equation if you want): $5(1)+4z=13$, so $4z=8$, so $z=2$. Lastly, put $x=1$ and $z=2$ back into one of the original equations, say the first one, to get the value for y: $1+y+2=4$, so $y+3=4$, so $y=1$. So the answer is $x=1$, $y=1$ and $z=4$.

Example: Solve the system

$$x + 2y - z = 1$$
$$x + y + z = 4$$
$$2x - y = 1$$

Solution: Notice that the third equation in this system has only two variables. Things still work, however, and, in fact, our job is easier. If we think of eliminating the z (the missing variable in equation 3), then we only have to work with one pair of equations (the first and second), rather than with two like in the previous example. So to eliminate the z, add the first and second equations together:

$$x + 2y - z = 1$$
$$\underline{x + y + z = 4}$$
$$2x + 3y = 5$$

So now we solve this system:

$$2x - y = 1$$
$$2x + 3y = 5$$

If we want to solve it by elimination, multiply the $2x - y = 1$ by 3 and add it to the other equation:

$$(2x - y = 1) \implies 6x - 3y = 3$$
$$2x + 3y = 5 \qquad \underline{2x + 3y = 5}$$
$$\qquad\qquad\qquad\qquad 8x = 8$$

So $x = 1$. Then we back substitute as before to get $y = 1$ and $z = 2$. So the solution is $x = 1$, $y = 1$, $z = 2$.

Much of the terminology we had for systems of two equations in two variables carries right over into larger systems. For example, we can still talk about a system being *consistent* if it has exactly one solution, *inconsistent* if it has no solutions and *dependent* if it has infinitely many solutions.

We can, of course, also use systems of three equations with three variables to solve story problems where we're looking for three different quantities. Mixing problems again show up, the only difference being that you're given three different pieces of information.

5.4 Linear Inequalities in Two Variables

What's Important
• Graph linear inequalities in two variables.

Earlier, we graphed inequalities in one variable on a number line. In this section, we'll graph (linear) inequalities that have two variables. Naturally, since there are two variables (mostly), we have to graph in two dimensions, on the standard coordinate plane. Our graphs will turn out to be regions (half-planes, to be specific) that have a straight line as a boundary.

So if someone hands you a linear inequality in two variables how do you graph it? Do the following two steps:

1. Graph the line corresponding to the inequality. Use a dotted line for strict inequalities, a solid line for nonstrict inequalities.

2. Decide which side of the line contains the points that satisfy the inequality and shade that side.

Example: Graph the inequality $x + y > 2$

Solution: The first step is to graph the line corresponding to the inequality. What is that line? To get it, just replace the greater than sign with an equals sign: $x + y = 2$. So we want to graph the line $x + y = 2$. Recall from Chapter 4 how to do this. We could put it in slope–intercept form (solve for y) then use the slope and y-intercept to graph it, or we could find the x-intercept (put 0 in for y and solve for x) and the y-intercept (put 0 in for x and solve for y), plot those points and draw the line. Let's try the second method here:

x-intercept: $x + 0 = 2$, so $x = 2$.

y-intercept: $0 + y = 2$, so $y = 2$.

So the x-intercept is $(2, 0)$ and the y-intercept is $(0, 2)$. So if we draw our line, we get this graph. (See Figure 5.2 on the top of the next page.) Note that we have a strict inequality (just *greater than*, not *greater than or equal to*) so we use a dotted line.

Next, we need to figure out which side of the line to shade. There are two ways to do this. One way is to pick a point on one side of the line (such as $(0, 0)$) and plug it into the inequality to see if we get a true statement. If we do, then we shade that side of the line, otherwise, we shade the other side. So if we try $(0, 0)$ in the inequality, we get $0 + 0 > 2$ or $0 > 2$ which is false, so we shade the other side of the line (the side that doesn't include $(0, 0)$).

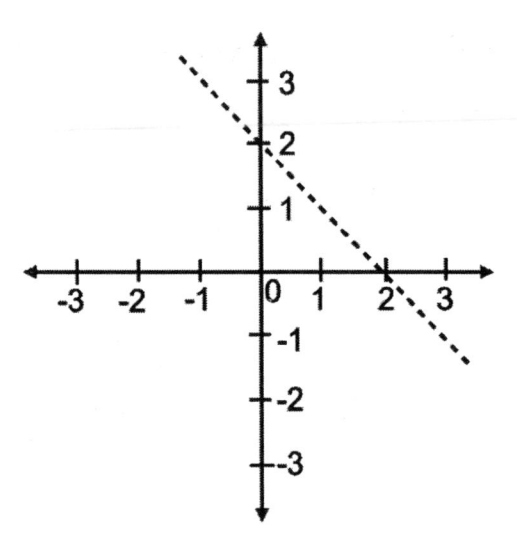

Figure 5.2: Graph of $x + y = 2$.

A second way to pick the correct side to shade is to look at the inequality and think along these lines: "Greater than" means "above" and "less than" means "below". So since we have a "greater than" inequality here, we must shade above the line (the same result as from the first method). I need to warn you about a couple of things for this method.

1. You always have to read the inequality from the point of view of the y. So an inequality like this $2 < x + y$ gets read "2 is less than $x + y$" (so you might think to shade below the line $x + y = 2$), but from the point of view of the y, it's really saying "$x + y$ is greater than 2" (the same as our example).

2. The sign of the y is important. If y is positive, then "greater than" does mean "above" and "less than" does mean "below", but if y is negative, then things get reversed: "Greater than" means "below" and "less than" means "above".

With all of that, you're probably better off testing a point (the first method), but I found that students seem to prefer the second method for some reason. Anyway, here's the final picture:

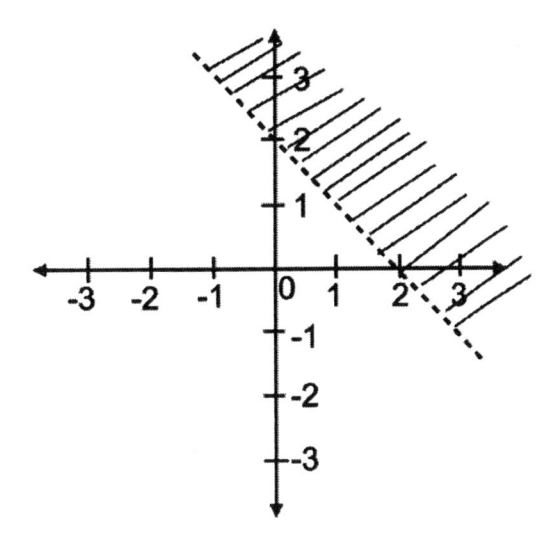

5.4. LINEAR INEQUALITIES IN TWO VARIABLES

Example: Graph the inequality $x \leq 2$

Solution: Notice that we only have one variable in this inequality (an x), but we'll still graph it on the coordinate plane. The boundary line will be $x = 2$, a vertical line. We'll use a solid line this time since we have a nonstrict inequality (less than or equal to). Next, if we pick a point (such as $(0,0)$) and put it in the inequality we get $0 \leq 2$, a true statement, so we shade the side of the line that contains the point $(0,0)$ (the left side). So our graph looks like this:

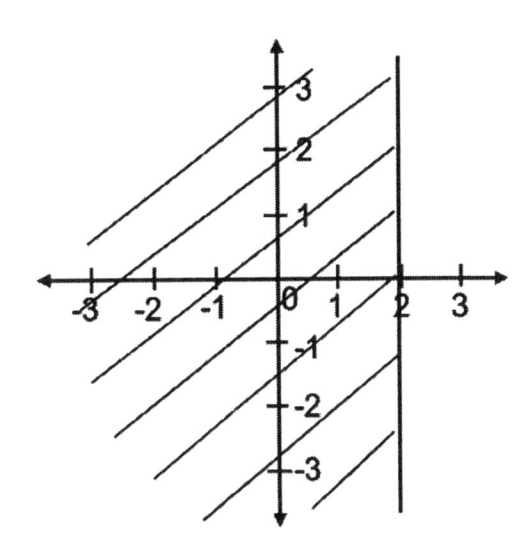

5.5 Systems of Linear Inequalities in Two Variables

What's Important

- Graph a system of linear inequalities.

And now we'll combine the two main topics of this chapter: systems and inequalities in two variables. So that means graphing systems of inequalities. The method is pretty much the same as for a single inequality: Draw the line corresponding to the inequality and shade the appropriate region. For a system, you have to draw several boundary lines and figure out which region satisfies all of the inequalities at the same time.

Example: Sketch the system:

$$x + y < 2$$
$$y - x > -2$$

Solution: We start by graphing the lines $x + y = 2$ and $y - x = -2$. Since we have strict inequalities, we'll use dotted lines:

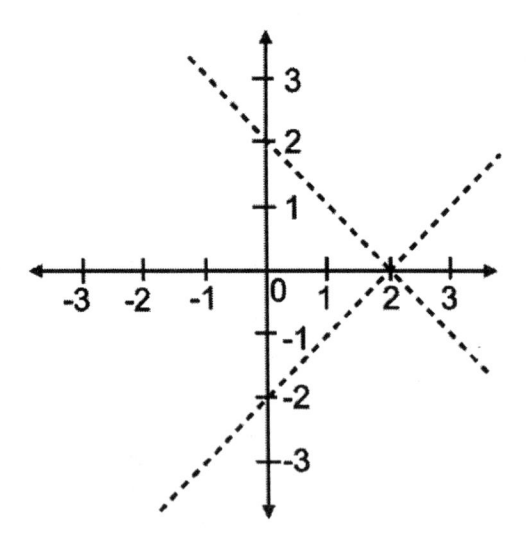

Next, we need to figure out the proper region to shade. Here we have a choice of four. The same two methods I mentioned in Section 5.4 will work: Choose a test point or look at the inequalities and think about above and below (but make sure you use them both at once). In this case, the region turns out to be above $y - x = -2$ and below $x + y = 2$, which is the

5.5. SYSTEMS OF LINEAR INEQUALITIES IN TWO VARIABLES

triangular region on the left. (If you used a test point, such as $(0,0)$, it turns out to be the region that contains $(0,0)$.) So the graph of the system looks like this:

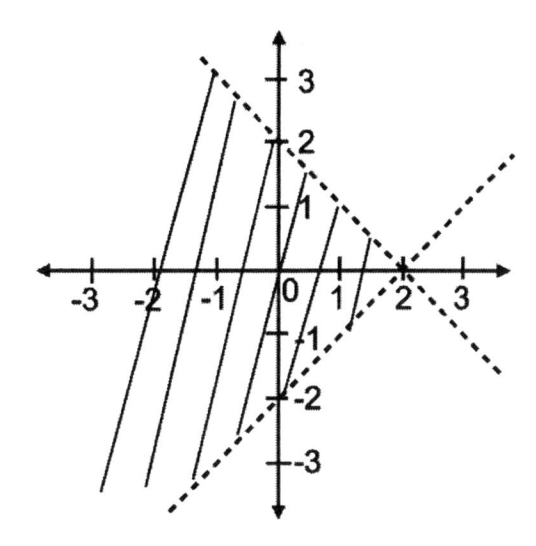

This kind of thing can be done with pretty much as many inequalities as you want (a real-world application of this stuff is called **linear programming**, and fairly simple problems may involve five or six inequalities), so let's try an example with four inequalities.

Example: Graph the system:

$$x + 3y < 3$$
$$y > x$$
$$x > -2$$
$$y > 0$$

Solution: We first plot the lines $x + 3y = 3$, $y = x$, $x = -2$ and $y = 0$. Then we figure out the region, which must be below $x + 3y = 3$, above $y = x$ to the right of $x = -2$ and above $y = 0$ (the x-axis). So the graph must look like the picture on the next page (Figure 5.3). If you didn't get this picture, try the test point $(-1, 1)$. This point works in all of the inequalities, so the region we shade must be the one that contains this point.

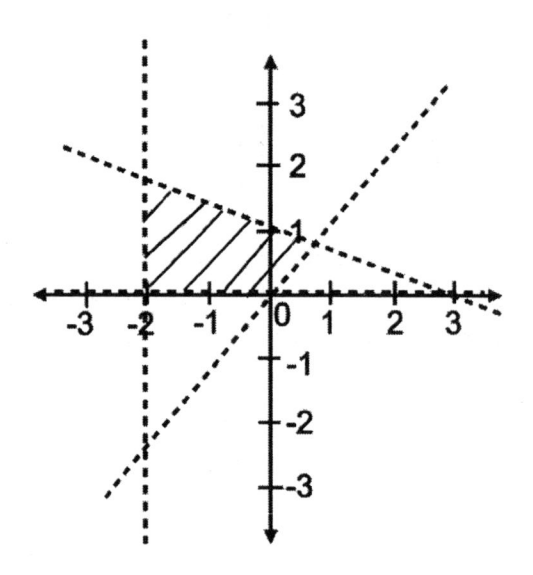

Figure 5.3: Graph of the system $x + 3y < 3, y > x, x > -2, y > 0$

Chapter 6

Polynomials

In this chapter, we take a look at polynomials. For most people, when they think of algebra, they think about working with polynomials – adding, subtracting, multiplying, or factoring them. So if you're like that, this is your chapter.

6.1 Polynomial ABCs

What's Important
- Identify like terms.
- Find the degree of a polynomial.

"Polynomial" is a word that means something like "many terms". So what's a term? For people in algebra, a **term** is a number (the **coefficient**) perhaps multiplying one or more letters (variables) with nonnegative integer exponents (meaning $0, 1, 2, \ldots$). So here are some terms:

$$4x, \ -5y^2w^{18}, \ m^{57}, \ 5$$

At this point, you might ask where's the coefficient of m^{57}? The answer is that if the coefficient is just 1 (one), we usually don't write it. For comparison, here are few things we won't call terms (if we're talking about polynomials):

$$\frac{5}{x}, \ 4x - 3y, \ 4\sqrt{w}$$

The reason these aren't terms are that in the first one we're dividing by x, in the second one we have subtraction thrown in there and in the third one we have the square root of a variable.

Now that we know what a term is, we can talk about like terms. **Like terms** have exactly the same letters raised to exactly the same powers (the coefficients can be different, though). For example, $4x$ and $-5x$ are like terms, but $4x$ and $-5y$ are not, since they have different variables. Similarly, $-5y^2w^{18}$ and $4y^2w^{18}$ are like terms, but $-5y^2w^{18}$ and $4y^3w^{18}$ are not, since though they have the same variables, the y has a different power in the first compared to the second.

The powers on the variables in a term are used to define its **degree**. To get the degree of a term, add all of the exponents together (in some books this called the **total degree**). So the degree of $4x$ is 1 (we just have x to the first power), the degree of $-5y^2w^{18}$ is $2 + 18 = 20$ and the degree of 5 is 0 (since there's no variable at all).

By adding and subtracting terms, we get polynomials. Some special polynomials you're likely to hear mentioned are:

- monomial – one term

- binomial – two terms

- trinomial – three terms

The degree of a polynomial is defined to be the degree of its highest degree term. For example, the degree of the trinomial $4x^2 - 8x + 5$ is 2 (the highest degree here is 2), and the degree of the binomial $-5y^2w^{18} + 4y^5w^3$ is 20 (the highest degree here is 20).

The last thing I'll mention in this section is a **polynomial function**, which is just a function that looks like a polynomial. Here are a couple of examples:

$$f(x) = 3x - 2$$
$$g(y) = 4y^2 - 8y + 7$$

For the moment, all we'll do with polynomial functions is admire them (aren't they pretty?) and evaluate them, but in the next section, we'll start combining them together to make new functions.

6.2 Combining Functions

> **What's Important**
> - Add, subtract, multiply and divide functions.

To add, subtract, multiply or divide functions, just follow these rules:

1. $(f+g)(x) = f(x) + g(x)$
2. $(f-g)(x) = f(x) - g(x)$
3. $(f \cdot g)(x) = f(x) \cdot g(x)$
4. $(f \div g)(x) = f(x) \div g(x)$, if $g(x) \neq 0$

In other words, combining functions is really about combining their values. For most problems though, you start with the formulas for the functions and combine them together.

Example: Let $f(x) = x+4$, $g(x) = x-5$. Find $(f+g)(x)$, $(f-g)(x)$, $(f \cdot g)(x)$, and $(f \div g)(x)$.

Solution:

1. $(f+g)(x) = (x+4) + (x-5) = 2x - 1$
2. $(f-g)(x) = (x+4) - (x-5) = 9$
3. $(f \cdot g)(x) = (x+4)(x-5) = x^2 - 5x + 4x - 20 = x^2 - x - 20$
4. $(f \div g)(x) = (x+4) \div (x-5) = \dfrac{x+4}{x-5}$

As with anything involving functions, you should always take a moment to think about domains (the numbers you can put into the function). For addition, subtraction and multiplication, the domain will always be all numbers that work in both functions. For division, the domain will be a little smaller. It will be all numbers that work in both functions, not including those that make the bottom (divisor) equal to zero. If you actually need to find the domain of a combination of functions, the best way to do it is to figure out the combined function first, then use what you know about finding domains on the combined function. For instance, the domain of $(f \div g)(x)$ from the example above will be $x \neq 5$, since 5 is the only number that doesn't work in the combined function $\dfrac{x+4}{x-5}$.

6.3 Adding and Subtracting Polynomials

> **What's Important**
> - Adding and subtracting polynomials.

In a more logically organized book, this section and the next one would come before Section 6.2, since in this section we look at how to actually add and subtract polynomials and in the next we look at how to actually multiply polynomials. But anyway, here goes: The basic rule for adding and subtracting polynomials is to get rid of parentheses and combine like terms. In subtraction, you must remember, of course, to carry the minus sign all the way through the second polynomial.

Example: Find $(4x^2 - 3x + 5) + (7x^2 + 5x - 2)$ and $(4x^2 - 3x + 5) - (7x^2 + 5x - 2)$.

Solution:

$$(4x^2 - 3x + 5) + (7x^2 + 5x - 2) = 4x^2 - 3x + 5 + 7x^2 + 5x - 2$$
$$= 11x^2 + 2x + 3$$
$$(4x^2 - 3x + 5) - (7x^2 + 5x - 2) = 4x^2 - 3x + 5 - 7x^2 - 5x + 2$$
$$= -3x^2 - 8x + 7$$

6.4 Multiplying Polynomials

> **What's Important**
> - Multiplying polynomials.

Multiplying polynomials makes use of several things:

- The Distributive Law
- Properties of exponents
- Combining like terms

It's traditional to start by multiplying monomials, then do a monomial times a polynomial, then a polynomial times a polynomial. Somewhere in there we usually also mention a binomial times

6.4. MULTIPLYING POLYNOMIALS

a binomial so we can mention FOILing. So to multiply monomials together, you *multiply* their coefficients and *add* exponents for the same letters.

Example: Multiply: $(5x^2y)(-3x^3y^3)$

Solution:

$$(5x^2y)(-3x^3y^3) = (5 \cdot (-3))(x^{2+3}y^{1+3})$$
$$= -15x^5y^4$$

To multiply a monomial and a polynomial, distribute it out first (which turns it into a combination of monomials multiplied together), do those individual multiplications and combine any like terms.

Example: Multiply: $5x^2(x^2 - 7x + 5)$

Solution:

$$5x^2(x^2 - 7x + 5) = (5x^2)(x^2) - (5x^2)(7x) + (5x^2)(5)$$
$$= 5x^4 - 35x^3 + 25x^2$$

A polynomial times a polynomial makes repeated use of the distributive law. The easiest example to start with is a binomial times a binomial. In this case, we talk about the FOIL method, where FOIL stands for First, Outside, Inside, Last, which gives the order in which you multiply terms. Notice that the distributive law gets used twice, once for each term in the first binomial.

Example: Multiply: $(x + 3)(x - 7)$

Solution:

$$(x + 3)(x - 7) = x^2 - 7x + 3x - 21$$
$$= x^2 - 4x - 21$$

For longer polynomials, it's the same pattern: multiply each term in the second polynomial by each term in the first polynomial and combine like terms.

Example: Multiply: $(x^2 + 4x - 3)(x^2 - 7x + 5)$

Solution:

$$(x^2 + 4x - 3)(x^2 - 7x + 5) = (x^2)(x^2) - (x^2)(7x) + (x^2)(5) + (4x)(x^2)$$
$$- (4x)(7x) + (4x)(5) - (3)(x^2)$$
$$+ (3)(7x) - (3)(5)$$
$$= x^4 - 7x^3 + 5x^2 + 4x^3 - 28x^2 + 20x - 3x^2$$
$$+ 21x - 15$$
$$= x^4 - 3x^3 - 26x^2 + 41x - 15$$

It's possible to multiply more than two polynomials together. To do this, multiply adjacent factors together, and keep multiplying until you run out of factors.

Example: Multiply: $(x + 3)(x + 2)(x + 1)$

Solution:

$$(x + 3)(x + 2)(x + 1) = (x + 3)(x^2 + x + 2x + 2)$$
$$= (x + 3)(x^2 + 3x + 2)$$
$$= x^3 + 3x^2 + 2x + 3x^2 + 9x + 6$$
$$= x^3 + 6x^2 + 11x + 6$$

Finally, all books mention a few special formulas for multiplying binomials together: "squaring binomials" and "difference of squares". Rather than learning these formulas, just get good at multiplying polynomials.

6.5 Starting to Factor

> What's Important
> - Take out common factors.
> - Factor by grouping.

In this last section we looked at multiplying polynomials. In this section we start to look at "unmultiplying" polynomials, or, as it's better known, **factoring polynomials**. Factoring polynomials is basically doing the distributive law backwards. The easiest kind of factoring (and always the first kind to try) is to take out common factors. So what's a common factor? In general, a common factor is anything that will divide into all of the terms in a polynomial. Much of the time it will look like a monomial; the coefficient will be a number that divides all of the coefficients of the polynomial and the variable part will have powers that are less than or equal to all of the powers (of that particular letter) that appear in the polynomial. When we take the common factor out, we divide by it.

Example: Factor $4x^3 - 8x^2 + 12x^5$

Solution: Notice that all of the coefficients here can be divided by 4, and all of the terms have a power of x in them. What's the biggest power of x that we can take out of all of them? It would be x^2. So the common factor for this polynomial would be $4x^2$ and if we factor it, we get
$$4x^3 - 8x^2 + 12x^5 = 4x^2(x - 2 + 3x^3)$$

Example: Factor $12x^3y^5 + 6x^5y^4 - 9x^4y^8$

Solution: For this polynomial, the common factor must be $3x^3y^4$ since 3 is the largest number that divides all of the coefficients, the highest power of x that goes into all of the terms is x^3 and the highest power of y that goes into all of the terms is y^4. So if we factor this we get
$$12x^3y^5 + 6x^5y^4 - 9x^4y^8 = 3x^3y^4(4y + 2x^2 - 3xy^4)$$

As the next example shows, the common factor doesn't have to be a monomial. This will lead into the next topic, **factoring by grouping**.

Example: Factor $x(x-3) + 5(x-3)$

Solution: In this polynomial, notice that both terms have a factor of $x - 3$. So $x - 3$ is the common factor! If we factor this out, we get
$$x(x-3) + 5(x-3) = (x-3)(x+5)$$

Factoring by grouping works on some polynomials that have an even number of terms (usually four terms). The idea is to divide the polynomial into two groups and take out the common factors from each group. Hopefully, we can then finish up by taking out a factor common to both groups.

Example: Factor $xa + xb + ya + yb$

Solution: Let's group together the first two terms and the second two terms:

$$xa + xb + ya + yb = (xa + xb) + (ya + yb)$$

Then x is a common factor for the first group, and y is a common factor for the second group, so we can factor them out:

$$(xa + xb) + (ya + yb) = x(a + b) + y(a + b)$$

And now notice that $a + b$ is a common factor of the two terms we have left. So just like in the last example, we factor it out and we're done

$$x(a + b) + y(a + b) = (a + b)(x + y)$$

6.6 Special Factoring Formulas

What's Important
- Factor polynomials using special formulas.

Certain types of polynomials show up in factoring problems so often that they have their own special formulas. (Some of these formulas were referred to in Section 6.4, though here they'll be written backwards.) The key to using these formulas is to first recognize the pattern, then to figure out how the formula applies to the actual polynomial.

1. Difference of squares: $A^2 - B^2 = (A - B)(A + B)$

2. Perfect square trinomial (addition): $A^2 + 2AB + B^2 = (A + B)^2$

3. Perfect square trinomial (subtraction): $A^2 - 2AB + B^2 = (A - B)^2$

4. Difference of cubes: $A^3 - B^3 = (A - B)(A^2 + AB + B^2)$

5. Sum of cubes: $A^3 + B^3 = (A + B)(A^2 - AB + B^2)$

6.7. FACTORING TRINOMIALS

Some comments: First, notice that I didn't give a formula for the sum of squares. That's because there isn't one, so don't try to come up with one. Second, the formulas for perfect square trinomials can be helpful, but once you know how to factor trinomials in general (see the next section) you won't need these formulas. Lastly, pay attention to the signs in the formulas for the sum of cubes and the difference of cubes. And now for some examples.

Example: Factor $x^2 - 49$

Solution: Notice that $49 = 7^2$, so we have the difference of squares here. To apply the formula, use $A = x$ and $B = 7$ to get

$$x^2 - 49 = (x - 7)(x + 7)$$

Example: Factor $x^2 - 14x + 49$

Solution: For this polynomial, notice that we have squares on the ends (x^2 and 7^2) and that in the middle we have something that looks like $2(7)(x)$. This matches the pattern for a perfect square trinomial, and the minus sign tells us to use formula 3 with $A = x$ and $B = 7$ to get

$$x^2 - 14x + 49 = (x - 7)^2$$

Example: Factor $x^3 + 8$

Solution: For this one, notice that $8 = 2^3$, so we must have the sum of cubes. To apply the formula, use $A = x$ and $B = 2$ and end up with

$$x^3 + 8 = (x + 2)(x^2 - 2x + 2^2) = (x + 2)(x^2 - 2x + 4)$$

6.7 Factoring Trinomials

What's Important

- Factoring trinomials.

Perhaps the most common type of factoring you'll run across in an algebra class is factoring a trinomial. As mentioned before, *trinomial* just means a polynomial with three terms, but actually, what we mean here is a polynomial that looks like this:

$$ax^2 + bx + c, \text{ where } a \neq 0$$

Another name for a polynomial like this is a **quadratic polynomial**. But for short, I'll still call it a trinomial.

In Section 6.4, we foiled two binomials together and got a trinomial:
$$(x+3)(x-7) = x^2 - 7x + 3x - 21 = x^2 - 4x - 21$$

In factoring a trinomial, we try to get two binomial factors. The middle step above gives us the clue on how to do this. The first terms in the factors must multiply together to give the first term in the trinomial. The last terms in the factors must multiply together to give the last term in the trinomial. The inside and outside products must add together to give the middle term of the polynomial. Sounds like a lot to keep track of, doesn't it? Unfortunately, it can be, especially for trinomials that have first and last terms with lots of factors. So in general, you may just have to play around with numbers until you find the right combination. And be sure to pay attention to signs! Here are a few rules about signs:

1. $ax^2 + bx + c$ – all plus signs in factors

2. $ax^2 - bx + c$ – last terms in factors have minus signs

3. $ax^2 - bx - c$ or $ax^2 + bx - c$ – one factor will be plus, the other minus.

Example: Factor $x^2 + 6x + 5$

Solution: This is the easiest type of trinomial, where there's no number in front of the x^2. We also have all plus signs here, so the basic pattern for factoring will look like

$$(x+)(x+)$$

To fill in the spaces, look at the factors of 5. In fact, the only factors of 5 are 1 and 5, and notice that $1 + 5 = 6$. So $x^2 + 6x + 5$ must factor as $(x+1)(x+5)$, and, in fact, if you multiply these together, you do get back to $x^2 + 6x + 5$. (You can always check that you factored correctly by multiplying the factors back together.)

Example: Factor $x^2 + 5x + 6$

Solution: This one still follow the pattern of $(x+)(x+)$, but there are more possibilities for the spaces than in the last example, since there are more ways to factor 6. The possible factors of 6 are 1 and 6 and 2 and 3. Which one do we choose? Remember, they need to add up to equal 5, so let's check things out. $1 + 6 = 7$, so 1 and 6 don't work, but $2 + 3 = 5$, so we must have $x^2 + 5x + 6 = (x+2)(x+3)$.

Example: Factor $x^2 - 4x - 12$

Solution: Notice that the last term here has a minus sign in front of it, so the factoring pattern must be $(x+)(x-)$. The factors of 12 are 1 and 12, 2 and 6, and 3 and 4. Again we have to check which pair adds up to -4, but in this case, one of them has to be negative and the other positive. It turns out that the correct choice is $2 + (-6) = -4$, so we have $x^2 - 4x - 12 = (x+2)(x-6)$.

Example: Factor $3x^2 + 7x - 6$

Solution: This is an example of the trickiest type of polynomial to factor, one where there is a number in front of the x^2 (3 in this case) and different signs in the factors (because of the minus sign in front of the 6). So to start this one, look at the factors of 3, which are just 3 and 1. So the factoring pattern will either be $(3x+)(x-)$ or $(3x-)(x+)$. Next, look at the factors of 6: 1 and 6, 2 and 3. Now we have to try out the different possibilities of signs and numbers so that we end up with $7x$ in the middle. You may find it helpful to actually write down your attempts:

$$(3x+1)(x-6) = 3x^2 - 17x - 6 \quad \text{Nope}$$
$$(3x-1)(x+6) = 3x^2 + 17x - 6 \quad \text{Nope}$$
$$(3x+6)(x-1) = 3x^2 + 3x - 6 \quad \text{Nope}$$
$$\vdots$$

If you keep trying, you eventually find it factors like this:

$$3x^2 + 7x - 6 = (3x - 2)(x + 3)$$

6.8 Dividing Polynomials

What's Important

- Divide a polynomial by a polynomial.

So far, we've looked at how to add, subtract, and multiply polynomials. Now we'll look at dividing them. First we'll look at dividing monomials. In a sense, dividing monomials is a lot like multiplying them. Recall that the way we multiplied them was that we multiplied their coefficients, then used properties of exponents to combine their variables. Well, we do the same kind of thing in dividing them. We divide their coefficients and use properties of exponents to combine their variables, in this case the property that when you divide things with exponents, you subtract the bottom exponent from the top.

Example: Divide: $\dfrac{12x^2y^4}{3xy^2}$

Solution: Divide the 3 into the 12 to get 4, and subtract the exponents on the x and y to get

$$\frac{12x^2y^4}{3xy^2} = 4x^{2-1}y^{4-2} = 4xy^2$$

The next step is to divide a polynomial by a monomial. To do this, you break up the problem into a combination of fractions, each of which is a monomial over a monomial and then divide as above.

Example: Divide: $\dfrac{12x^2y^4 + 18x^7y^{10}}{3xy^2}$

Solution: So we break up into two fractions:

$$\frac{12x^2y^4 + 18x^7y^{10}}{3xy^2} = \frac{12x^2y^4}{3xy^2} + \frac{18x^7y^{10}}{3xy^2}$$

and divide

$$\frac{12x^2y^4}{3xy^2} + \frac{18x^7y^{10}}{3xy^2} = 4x^{2-1}y^{4-2} + 6x^{7-1}y^{10-2}$$
$$= 4xy^2 + 6x^6y^8$$

Finally, we get to dividing a polynomial by another polynomial. To do this, we have to use a different method: long division. Recall how long division works with numbers.

Example: $17\overline{)439}$

Solution: You say: 17 goes into 43 how many times? 2 times, so you write a 2 on top (over the 3), multiply the 17 by 2 to get 34, subtract 34 from 43, and bring down the next number.

$$\begin{array}{r} 2 \\ 17\overline{)439} \\ 34 \\ \hline 99 \end{array}$$

And then you figure out 17 goes into 99 5 times, you write the 5 on top (over the 9), multiply the 17 by 5 to get 85 and subtract the 85 from the 99.

6.8. DIVIDING POLYNOMIALS

$$\begin{array}{r} 25 \\ 17{\overline{\smash{\big)}\,439}} \\ \underline{34} \\ 99 \\ \underline{85} \\ 14 \end{array}$$

At this point, you stop, since 14 is smaller than 17. So your **quotient** is 25 and your **remainder** is 14.

Long division of polynomials works much like this. The main differences are that

1. You always try to divide the first term of the divisor into the first term of the remainder.

2. You stop when the *degree* of the remainder is less than the degree of the divisor.

Example: Divide: $x+2\overline{)x^2 - 4x + 3}$

Solution: Start by dividing x into x^2 (the first term of the divisor $x+2$ into the first term of $x^2 - 4x + 3$), or say to yourself "What times x equals x^2?" The answer is x, so write an x above the x^2 and multiply $x+2$ by x to get $x^2 + 2x$:

$$\begin{array}{r} x \\ x+2{\overline{\smash{\big)}\,x^2 - 4x + 3}} \\ \underline{x^2 + 2x} \end{array}$$

Now subtract: $x^2 - x^2 = 0$ and $-4x - 2x = -6x$ and bring down the next term:

$$\begin{array}{r} x \\ x+2{\overline{\smash{\big)}\,x^2 - 4x + 3}} \\ \underline{x^2 + 2x} \\ -6x + 3 \end{array}$$

Now repeat the process, so what times x equals $-6x$? Write a -6 on top, multiply $x+2$ by -6 to get $-6x - 12$ and subtract:

$$\begin{array}{r} x - 6 \\ x+2{\overline{\smash{\big)}\,x^2 - 4x + 3}} \\ \underline{x^2 + 2x} \\ -6x + 3 \\ \underline{-6x - 12} \\ 15 \end{array}$$

Did you remember to pay attention to signs? So $-6x-(-6x)=0$ and $3-(-12)=15$. Anyway, the degree of the polynomial 15 is less than the degree of $x+2$, so we're done. So we can say that the quotient is $x-6$ and the remainder is 15. (Another way to write this is in the form

$$\frac{x^2-4x+3}{x+2}=x-6+\frac{15}{x+2},$$

but I prefer writing things separately.)

As a last example, we'll look at a division problem where the polynomial you're dividing into is "missing" a term.

Example: Divide: $x-3\overline{)x^3-4x+7}$

Solution: To do this division, start by rewriting x^3-4x+7 as x^3+0x^2-4x+7. Then just divide as above.

$$\begin{array}{r} x^2+3x+5 \\ x-3{\overline{\smash{\big)}\,x^3+0x^2-4x+7}} \\ \underline{x^3-3x^2} \\ 3x^2-4x \\ \underline{3x^2-9x} \\ 5x+7 \\ \underline{5x-15} \\ 22 \end{array}$$

So the quotient is x^2+3x+5 and the remainder is 22.

6.9 Solving Equations by Factoring

What's Important

- Solve equations by factoring.

A common use for factoring trinomials is in solving **quadratic equations**. A quadratic equation is any equation that can be written like this:

$$ax^2+bx+c=0, a\neq 0$$

For example, $x^2-7x+10=0$ is a quadratic equation, as is $x^2=5$ (just think of moving the 5 to the left side), but $4x-3=0$ is not a quadratic equation (no x^2 in it).

6.9. SOLVING EQUATIONS BY FACTORING

To solve a quadratic equation by factoring, we use what most books call the *Zero Product Rule*, which says that if you multiply two things together and get zero, then at least one of the things you multiplied together must be zero. It may be clearer in symbols:

$$\text{If } a \cdot b = 0, \text{ then } a = 0 \text{ or } b = 0$$

So here's how it works:

Example: Solve $x^2 - 7x + 10 = 0$

Solution: Start by factoring the left side to get:

$$(x-5)(x-2) = 0$$

Notice that we have two things multiplied together to get zero. So by the rule, we must have $x - 5 = 0$ or $x - 2 = 0$. Now these are nice linear equations that we can solve by getting the x by itself. So if we solve the first one we get $x = 5$ and if we solve the second one, we get $x = 2$. So the solutions are $x = 5$ and $x = 2$.

Example: Solve $x^2 = 7x$

Solution: For this problem, the first step is to get zero on one side. So move the $7x$ over to the left side:

$$x^2 - 7x = 0$$

Now factor to get

$$x(x-7) = 0$$

Set each factor equal to zero

$$x = 0 \text{ or } x - 7 = 0$$

and solve to get $x = 0$ and $x = 7$ as solutions.

Note that this technique works for longer polynomials too. Factor the polynomial, then set each factor equal to zero and solve.

Example: Solve $x^3 + x^2 - 4x - 4 = 0$

Solution: Remember "factoring by grouping" from Section 6.5? That's what we'll use to factor this polynomial.

$$x^3 + x^2 - 4x - 4 = 0$$
$$(x^3 + x^2) - (4x + 4) = 0$$
$$x^2(x + 1) - 4(x + 1) = 0$$
$$(x + 1)(x^2 - 4) = 0$$

And now the $x^2 - 4$ is the difference of two squares, so if we use the special formula for that (see Section 6.6), we end up with

$$(x + 1)(x + 2)(x - 2) = 0$$

Set the factors equal to zero and solve for x and we end up with $x = -1$ or $x = -2$ or $x = 2$ as solutions.

Chapter 7

Rational Expressions

The next algebraic object to look at is a **rational expression**. In its simplest form, it's a fraction made from two polynomials. But, of course, things can get more complicated. In this chapter, we'll use what we know about exponents, factoring and number fractions to work with rational expressions.

7.1 Simplifying Rational Expressions

What's Important

- Simplify rational expressions

As mentioned above, a basic rational expression is a fraction made from two polynomials. So, for example, $\frac{x^2-4}{x+1}$ is a rational expression, as is $\frac{1}{x^2-7x+10}$. For any rational expression, we always assume the bottom doesn't equal zero (can't divide by zero, of course), so it's usually important to figure out what numbers would cause problems. To figure this out, set the bottom equal to zero and solve.

Example: Where is $\frac{x^2-4}{x+1}$ undefined?

Solution: To answer this question, we'll solve $x+1=0$. Get the x by itself, and you end up with $x=-1$. So $\frac{x^2-4}{x+1}$ is undefined when $x=-1$.

Simplifying a rational expression is just like simplifying a number fraction (reducing it to lowest terms): Take out the common factors from the top and bottom. But let me warn you that you can only cancel out things that are *multiplied*. You can't cancel out things that are added or subtracted.

Example: Simplify $\dfrac{4x^2y^7}{2x^4y^3}$

Solution: Notice that you can divide top and bottom by 2, you can divide top and bottom by x^2 and you can divide top and bottom by y^3. If you actually do these divisions, you're left with

$$\frac{4x^2y^7}{2x^4y^3} = \frac{2y^{7-3}}{x^{4-2}} = \frac{2y^4}{x^2}$$

Example: Simplify $\dfrac{x^2 - 4}{x^2 - 7x + 10}$

Solution: To simplify this expression, start by factoring the top and the bottom:

$$\frac{x^2-4}{x^2-7x+10} = \frac{(x+2)(x-2)}{(x-2)(x-5)}$$

Notice that you have an $x-2$ on the top and the bottom, so we'll divide that out and we end up with

$$\frac{(x+2)\cancel{(x-2)}}{\cancel{(x-2)}(x-5)} = \frac{x+2}{x-5}$$

In regards to my warning above, notice that the things we're left with are all added or subtracted, so we can't, for instance, cancel out the x's.

We can also talk about **rational functions**, which are functions whose formula is a rational expression. For example, $f(x) = \dfrac{x^2-4}{x+1}$ is a rational function, as is $f(x) = x^2 - 4$. (It may seem confusing, but all polynomial functions are rational functions too.) For rational functions, its very important to know where they're defined (their domain), which is one reason why we looked at how to figure out where a rational expression is not defined. For example, $f(x) = \dfrac{x^2-4}{x+1}$ is not defined at $x = -1$, so we would say its domain is $x \neq -1$.

It turns out that it's possible to simplify some rational functions (do it the same way as you simplify a rational expression), but even when it's simplified, you have to keep the same domain as the unsimplified version.

7.1. SIMPLIFYING RATIONAL EXPRESSIONS

Example: Simplify $f(x) = \dfrac{x^2 - 4}{x^2 - 7x + 10}$

Solution: To simplify this, start by factoring the top and bottom:

$$f(x) = \frac{x^2 - 4}{x^2 - 7x + 10} = \frac{(x-2)(x+2)}{(x-2)(x-5)}$$

but stop for a moment. Notice that if you set the bottom equal to zero and solve, you get $x = 2$ or $x = 5$. So the domain of f is $x \neq 2, 5$. Now go ahead and divide out the factor of $x - 2$ from the top and bottom to get

$$f(x) = \frac{\cancel{(x-2)}(x+2)}{\cancel{(x-2)}(x-5)} = \frac{x+2}{x-5}$$

Now it looks like the domain should be just $x \neq 5$, but we have to keep the old domain. So the proper way to write the simplified version is:

$$f(x) = \frac{x+2}{x-5}, \; x \neq 2, 5$$

A last question we might ask about a rational function is "What does its graph look like?" For most rational functions, we can't answer this question in this course, but for ones that have a special form, we can draw the graph.

Example: Graph $f(x) = \dfrac{x^2 + 3x + 2}{x + 1}$

Solution: Note that this function is not defined at $x = -1$. But also note that we can simplify it:

$$f(x) = \frac{x^2 + 3x + 2}{x + 1} = \frac{(x+2)\cancel{(x+1)}}{\cancel{x+1}} = x + 2$$

So we can write this function as $f(x) = x + 2$, $x \neq -1$. Now what's the graph look like? Well, the graph of $y = x + 2$ is a straight line with slope 1 and y-intercept $(0, 2)$, but what about the $x \neq -1$? It turns out that this means the line has a hole in it at $x = -1$. To figure out the exact coordinates of the point, put -1 into the simplified version of $f(x)$: $f(-1) = -1 + 2 = 1$. Then to sketch the graph, draw the line $y = x + 2$, but put a hole in the line at the point $(-1, 1)$. So the graph looks like this:

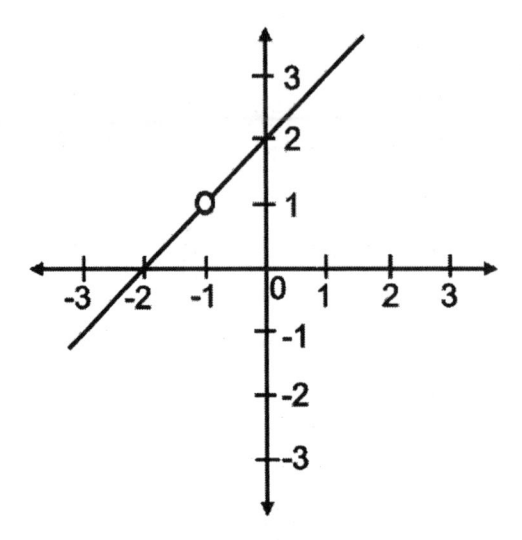

7.2 Multiplying and Dividing Rational Expressions

> **What's Important**
>
> - Multiply and divide rational expressions.

Multiplying and dividing rational expressions is just like multiplying and dividing number fractions:

- You multiply straight across (top times top, bottom times bottom).
- You divide by flipping the second fraction over and multiplying ("invert and multiply").

Simplifying also gets involved either at the very end of the calculation or in the middle. (In the middle will turn out to be the best place when you're working with rational expressions.)

Example: Multiply: $\dfrac{3x^2y^3}{2ab} \cdot \dfrac{4a^2b^5}{3xy}$

Solution: We'll just multiply straight across:

$$\frac{3x^2y^3}{2ab} \cdot \frac{4a^2b^5}{3xy} = \frac{12a^2b^5x^2y^3}{6abxy}$$

and then simplify:

$$\frac{12a^2b^5x^2y^3}{6abxy} = \frac{2ab^4xy^2}{1} = 2ab^4xy^2$$

7.2. MULTIPLYING AND DIVIDING RATIONAL EXPRESSIONS

In this next example, we'll simplify as we go along.

Example: Multiply: $\dfrac{x^2 - 4}{x^2 - 9} \cdot \dfrac{x + 3}{x^2 - 7x + 10}$

Solution: For this problem, we'll start by factoring everything:

$$\frac{x^2 - 4}{x^2 - 9} \cdot \frac{x + 3}{x^2 - 7x + 10} = \frac{(x+2)(x-2)}{(x+3)(x-3)} \cdot \frac{x + 3}{(x - 5)(x - 2)}$$

and then we'll cancel out the common factors from the top and bottom (in this case, the $x + 3$ and the $x - 2$) and multiply:

$$\frac{(x+2)\cancel{(x-2)}}{\cancel{(x+3)}(x-3)} \cdot \frac{\cancel{x+3}}{(x-5)\cancel{(x-2)}} = \frac{x+2}{x-3} \cdot \frac{1}{x-5} = \frac{x+2}{(x-3)(x-5)}$$

We'll finish up with a division problem.

Example: Divide: $\dfrac{x^2 - 7x + 10}{x^2 - 9} \div \dfrac{x^2 - 25}{x^2 - 4x + 3}$

Solution: Start by factoring and flipping the second fraction:

$$\frac{x^2 - 7x + 10}{x^2 - 9} \div \frac{x^2 - 25}{x^2 - 4x + 3} = \frac{(x-5)(x-2)}{(x-3)(x+3)} \cdot \frac{(x-3)(x-1)}{(x-5)(x+5)}$$

Then cancel out common factors from the top and bottom and multiply:

$$\frac{\cancel{(x-5)}(x-2)}{\cancel{(x-3)}(x+3)} \cdot \frac{\cancel{(x-3)}(x-1)}{\cancel{(x-5)}(x+5)} = \frac{(x-2)(x-1)}{(x+3)(x+5)}$$

7.3 Adding and Subtracting Rational Expressions

> **What's Important**
> - Find the least common denominator of two or more rational expressions.
> - Add and subtract rational expressions.

Remember that if you add and subtract number fractions that have the same denominator, you just add or subtract the numerators:

$$\frac{7}{5} + \frac{4}{5} = \frac{7+4}{5} = \frac{11}{5}$$

$$\frac{7}{5} - \frac{4}{5} = \frac{7-4}{5} = \frac{11}{5}$$

The same rule holds for rational expressions: If the denominators are the same, just add or subtract the tops. (Watch your signs!)

Example: Add: $\dfrac{x+3}{x-2} + \dfrac{x-7}{x-2}$

Solution: For this problem, the denominators are the same, so we just add the tops:

$$\frac{x+3}{x-2} + \frac{x-7}{x-2} = \frac{(x+3)+(x-7)}{x-2} = \frac{2x-4}{x-2}$$

Notice that it's possible to simplify this further:

$$\frac{2x-4}{x-2} = \frac{2(x-2)}{x-2} = \frac{2}{1} = 2$$

So the final answer is just 2.

What about when the denominators are different? For number fractions, you had to first find a common denominator. For example, the least common denominator for the fractions $\dfrac{2}{15}$ and $\dfrac{7}{25}$ is 75, because 75 is the smallest number that both 15 and 25 divide. Then once you had your LCD, you found equivalent fractions that had the LCD by multiplying top and bottom of your original fractions by some number. In this example, you would multiply the first fraction top and bottom by 5, since $75 \div 15 = 5$ and you would multiply the second fraction top and

7.3. ADDING AND SUBTRACTING RATIONAL EXPRESSIONS

bottom by 3, since $75 \div 25 = 3$. So the equivalent fractions would be

$$\frac{2 \cdot 5}{15 \cdot 5} = \frac{10}{75} \quad \text{and} \quad \frac{7 \cdot 3}{25 \cdot 3} = \frac{21}{75}$$

Rational expressions work exactly the same way (though more work is involved). To find the LCD for two (or more) rational expressions, factor their denominators completely, then find the smallest number that the coefficients will both divide and for each distinct factor, pick the highest power of that factor that appears in any of the denominators.

Example: Find the LCD for $\dfrac{5}{8x^3y^5}$ and $\dfrac{1}{3x^4y}$

Solution: In this example, everything is factored already, so we can start figuring the LCD out right away. First, the smallest number that both 8 and 3 divide is 24, so our LCD will have a coefficient of 24. Second, the highest power of x that appears in a denominator is x^4, so our LCD will have an x^4 in it. Finally, the highest power of y that appears in a denominator is y^5, so our LCD will have a y^5 in it. So the actual LCD is $24x^4y^5$.

What would the equivalent rational expressions that have a denominator of $24x^4y^5$ be? Well, we would have to multiply the first one top and bottom by $3x$, since $24x^4y^5 \div 8x^3y^5 = 3x$ and we would multiply the second one top and bottom by $8y^4$, since $24x^4y^5 \div 3x^4y = 8y^4$. So the equivalent expressions would be:

$$\frac{5 \cdot 3x}{8x^3y^5 \cdot 3x} = \frac{15x}{24x^4y^5} \quad \text{and} \quad \frac{1 \cdot 8y^4}{3x^4y \cdot 8y^4} = \frac{8y^4}{24x^4y^5}$$

Example: Find the LCD for $\dfrac{x+2}{x-3}$ and $\dfrac{x-7}{x^2-9}$

Solution: For this example, the first thing we need to do is to factor the bottom of the second expression:

$$\frac{x-7}{x^2-9} = \frac{x-7}{(x+3)(x-3)}$$

And now if we look at the different factors, both denominators have an $x - 3$, so that will be part of the LCD, and the second denominator has a factor of $x + 3$, so that will also be part of the LCD. So the LCD must be $(x+3)(x-3)$.

What are the equivalent rational expressions? For the first one, we need to multiply top and bottom by $x + 3$, since that's the missing factor, while for the second one, it already has the

LCD, so we don't need to do anything to it. So the equivalent expressions are

$$\frac{(x+2) \cdot (x+3)}{(x-3) \cdot (x+3)} = \frac{(x+2)(x+3)}{(x-3)(x+3)} \quad \text{and} \quad \frac{x-7}{(x-3)(x+3)}$$

And so to add or subtract rational expressions with different denominators

1. Find the least common denominator.

2. Find the equivalent expressions that have that LCD.

3. Add or subtract the tops of those expressions.

Example: Subtract: $\dfrac{x+2}{x-3} - \dfrac{x-7}{x^2-9}$

Solution: These are the same rational expressions that we looked at in the last example, so we know the LCD is $(x-3)(x+3)$ and the equivalent expressions are $\dfrac{(x+2)(x+3)}{(x-3)(x+3)}$ and $\dfrac{x-7}{(x-3)(x+3)}$. So we subtract the tops:

$$\begin{aligned}
\frac{x+2}{x-3} - \frac{x-7}{x^2-9} &= \frac{(x+2)(x+3)}{(x-3)(x+3)} - \frac{x-7}{(x-3)(x+3)} \\
&= \frac{(x+2)(x+3) - (x-7)}{(x-3)(x+3)} \\
&= \frac{x^2 + 5x + 6 - x + 7}{(x-3)(x+3)} \\
&= \frac{x^2 + 4x + 13}{(x-3)(x+3)}
\end{aligned}$$

7.4 Complex Fractions

> What's Important
>
> - Simplify complex fractions.

A more complicated–looking expression than what we've seen so far is called a **complex fraction**. A complex fraction is a rational expression that's built out of other rational expressions, meaning it will have fractions on the top or bottom (or both). Here are some examples:

$$\frac{\frac{1}{2}}{\frac{5}{3}} \qquad \frac{4}{1+\frac{x}{x+3}} \qquad \frac{1+\frac{1}{x}}{1-\frac{1}{x}}$$

There are two standard ways to simplify complex fractions:

1. Find the LCD of all of the fractions on the top and the bottom, multiply the whole complex fraction top and bottom by the LCD, then simplify more if necessary.

2. Add or subtract (as the case may be) the expressions on top and bottom separately (as in Section 7.3), then divide the top by the bottom (as in Section 7.2).

My experience is that students like the second method better, so I'll use it in the examples. But feel free to use the other method if you prefer.

Example: Simplify $\dfrac{\frac{1}{2}}{\frac{5}{3}}$

Solution: For this expression, we don't need to do any addition or subtraction on the top or bottom, so we can just divide (by inverting and multiplying):

$$\frac{\frac{1}{2}}{\frac{5}{3}} = \frac{1}{2} \cdot \frac{3}{5} = \frac{3}{10}$$

Example: Simplify $\dfrac{4}{1+\frac{x}{x+3}}$

Solution: We have to start by adding together the things on the bottom. Treat the 1 as the fraction $\dfrac{1}{1}$, so the LCD for the bottom will be $x+3$. So if we work on the bottom we get

$$\frac{4}{1+\frac{x}{x+3}} = \frac{4}{\frac{x+3}{x+3}+\frac{x}{x+3}} = \frac{4}{\frac{x+3+x}{x+3}} = \frac{4}{\frac{2x+3}{x+3}}$$

Then we invert and multiply to get

$$\frac{4}{\frac{2x+3}{x+3}} = \frac{4}{1} \cdot \frac{x+3}{2x+3} = \frac{4(x+3)}{2x+3}$$

Example: Simplify $\dfrac{1+\frac{1}{x}}{1-\frac{1}{x^2}}$

Solution: For this problem, we have to start by combining things together on the top and the bottom. The LCD for the top will be just x, while the LCD for the bottom will be x^2. So if we do the combining we get

$$\frac{1+\frac{1}{x}}{1-\frac{1}{x^2}} = \frac{\frac{x}{x}+\frac{x+1}{x}}{\frac{x^2}{x^2}-\frac{1}{x^2}} = \frac{\frac{x+1}{x}}{\frac{x^2-1}{x^2}}$$

Then we invert and multiply and simplify:

$$\frac{\frac{x+1}{x}}{\frac{x^2-1}{x^2}} = \frac{x+1}{x} \cdot \frac{x^2}{x^2-1}$$

$$= \frac{\cancel{x+1}}{\cancel{x}} \cdot \frac{\cancel{x^2}^{x}}{\cancel{(x+1)}(x-1)}$$

$$= \frac{x}{x-1}$$

7.5 Solving Rational Equations

> **What's Important**
>
> - Solve equations involving rational expressions.

A **rational equation** is an equation that contains rational expressions. A simple example:

$$\frac{4x}{3} + \frac{5}{6} = 8$$

A more complicated example:

$$\frac{4}{x-1} + \frac{4}{x+2} = 3$$

No matter how complicated the rational equation, they all get solved in the same way:

1. Multiply by the LCD of all of the terms to clear away the denominators.

2. Solve the (polynomial) equation that's left.

3. Check to make sure your answers don't make part of your original equation undefined.

That last step is very important to remember, since it's possible to get answers from step 2 that won't work in the original equation.

Example: Solve $\frac{4x}{3} + \frac{5}{6} = 8$

Solution: Start by figuring out the LCD, which turns out to be 6. Then we'll multiply *everything* in the equation by 6 and simplify:

$$6 \cdot \frac{4x}{3} + 6 \cdot \frac{5}{6} = 6 \cdot 8$$
$$\frac{24x}{3} + \frac{30}{6} = 48$$
$$8x + 5 = 48$$

And now if we solve for x, we get $8x = 43$, so $x = 43/8$. As for step 3, all of the denominators in our original equation were numbers, so it's never undefined. So the answer is $x = 43/8$.

Example: Solve $\dfrac{4}{x-1} + \dfrac{4}{x+2} = 5$

Solution: The LCD for this equation is $(x-1)(x+2)$, so we'll multiply everything by it and simplify.

$$\dfrac{4 \cdot \cancel{(x-1)}(x+2)}{\cancel{x-1}} + \dfrac{4 \cdot (x-1)\cancel{(x+2)}}{\cancel{x+2}} = 5 \cdot (x-1)(x+2)$$
$$4(x+2) + 4(x-1) = 5(x-1)(x+2)$$
$$4x + 8 + 4x - 4 = 5x^2 + 5x - 10$$
$$8x + 4 = 5x^2 + 5x - 10$$

Now if we move everything to the right side, we'll have a quadratic equation that we can solve by factoring (see Section 6.9 if you don't remember how).

$$0 = 5x^2 - 3x - 14$$
$$0 = (5x + 7)(x - 2)$$

Now set each factor equal to zero and solve:

$$5x + 7 = 0 \qquad\qquad x - 2 = 0$$
$$5x = -7 \qquad\qquad x = 2$$
$$x = -\dfrac{7}{5}$$

If we check $x = -7/5$ and $x = 2$ in the original equation, neither make it undefined, so these are the answers.

Example: Solve $\dfrac{x-2}{x-3} = \dfrac{3}{x^2 - 3x}$

Solution: To find the LCD for this equation, you have to factor the $x^2 - 3x$ first:

$$x^2 - 3x = x(x-3)$$

7.6. STORY PROBLEMS AND RATIONAL EQUATIONS

So the LCD must be $x(x-3)$, so let's multiply everything by it and simplify:

$$\frac{x\cancel{(x-3)} \cdot (x-2)}{\cancel{x-3}} = \frac{3 \cdot \cancel{x(x-3)}}{\cancel{x(x-3)}}$$
$$x(x-2) = 3$$
$$x^2 - 2x = 3$$
$$x^2 - 2x - 3 = 0$$

This factors as $(x-3)(x+1) = 0$, so we get $x = 3$ and $x = -1$. Notice what happens when we try these in the original equation. There's no problem with $x = -1$, but $x = 3$ makes the expressions undefined (zero on the bottom!). So we have to throw out that answer, and say the only solution is $x = -1$.

7.6 Story Problems and Rational Equations

> **What's Important**
> - Solve story problems using rational equations.

Actually, there's more to this section than just story problems, though they're the important part. First, literal equations (remember Section 2.2?) make another appearance, in this case, literal rational equations.

Example: Solve $\dfrac{1}{w} + \dfrac{2}{y} = \dfrac{3}{x}$ for w.

Solution: We still have a rational equation, so we still start by figuring out the LCD, which happens to be wxy. So let's multiply everything by wxy and simplify.

$$\frac{\cancel{w}xy \cdot 1}{\cancel{w}} + \frac{wx\cancel{y} \cdot 2}{\cancel{y}} = \frac{w\cancel{x}y \cdot 3}{\cancel{x}}$$
$$xy + 2wx = 3wy$$

Now remember to solve this, we need to get all of the w's on one side and isolate the w. So

$$xy = 3wy - 2wx$$
$$xy = w(3y - 2x)$$
$$\frac{xy}{3y - 2x} = \frac{w\cancel{(3y - 2x)}}{\cancel{3y - 2x}}$$
$$\frac{xy}{3y - 2x} = w$$

So we get $w = \dfrac{xy}{3y - 2x}$.

There are two main kinds of story problems where rational equations show up:

1. Motion problems (we've actually seen some before in Section 2.3)

2. Work problems

Example: A man rows a canoe 6 miles upstream against a 2 mph current, then rows back downstream to his starting point. If the whole trip took 4 hours, how fast can he row in still water?

Solution: Remember when we solved problems like this before, we picked a variable to represent the unknown (we'll let x represent his speed in still water) and made a table to organize the information.

	Rate	Time	Distance
Downstream	$x + 2$	$\dfrac{6}{x+2}$	6
Upstream	$x - 2$	$\dfrac{6}{x-2}$	6

At this point, you might ask where the entries came from. The entries for distance are obvious; it was six miles both ways. The entries for rate come from the fact that going downstream, the current helps so the speed of the current gets added to the boat speed, while going upstream, the current hinders so the speed of the current gets subtracted from the boat speed. For the time entries, we go back to the equation that relates distance, rate and time ($d = rt$) and solve it for time ($t = \dfrac{d}{r}$).

7.6. STORY PROBLEMS AND RATIONAL EQUATIONS

Now to get the equation, we know the total time for the trip was 4 hours, so we add the two entries for time together and set that equal to 4:

$$\frac{6}{x+2} + \frac{6}{x-2} = 4$$

And now solve this the same way we solved rational equations in the last section. Multiply by the LCD $((x+2)(x-2))$, simplify, and solve:

$$\frac{\cancel{(x+2)}(x-2) \cdot 6}{\cancel{x+2}} + \frac{(x+2)\cancel{(x-2)} \cdot 6}{\cancel{x-2}} = 4 \cdot (x+2)(x-2)$$
$$6(x+2) + 6(x-2) = 4(x+2)(x-2)$$
$$6x + 12 + 6x - 12 = 4x^2 - 16$$
$$12x = 4x^2 - 16$$
$$0 = 4x^2 - 12x - 16$$
$$0 = 4(x-4)(x+1)$$

So we get $x = 4$ or $x = -1$. We don't find any problems if we check these in the original equation, but a speed of -1 mph doesn't make sense, so we'll throw it out. So the answer is that he can row 4 mph in still water.

Work problems are the other type we'll look at. These are those problems where one person can do a job in so many hours, a second person can do the same job in a different number of hours, so how long will it take them to do it together. (Stephen Leacock, a Canadian humorist, parodied these problems in a funny short story once.) The basic idea used in these problems is that if you can do a job in 3 hours, then you'll do one-third of the job in one hour. (Streeter, et al. calls this the *Work Principle I.*) In general, all of the terms in your work equation will look like this, and you'll just be adding the individual contributions to get the total work. (We'll stick with two people, but you can add more terms if you have more employees.)

Example: John can mow the lawn in 5 hours. Marsha can mow the lawn in 3 hours. How long would it take them to mow the lawn together?

Solution: We'll let x stand for how long it will take the two of them to mow the lawn together. As mentioned above, we're going to add the individual contributions to get the total work. John does 1/5 of the lawn in 1 hour, Marsha does 1/3 of the lawn in 1 hours, and $1/x$ of the lawn gets done by the two of them in one hour. So the equation must look like:

$$\frac{1}{5} + \frac{1}{3} = \frac{1}{x}$$

And now solve this in the usual way: Multiply by $15x$ (the LCD), simplify and solve.

$$\frac{1 \cdot 15x}{5} + \frac{1 \cdot 15x}{3} = \frac{1 \cdot 15\cancel{x}}{\cancel{x}}$$
$$3x + 5x = 15$$
$$8x = 15$$
$$x = \frac{15}{8}$$

So it takes them 15/8 hours or $1\frac{7}{8}$ hours or 1 hour, 52.5 minutes to mow the lawn together.

7.7 Negative Exponents

> What's Important
> - Work with negative exponents.

We'll actually start out by defining zero as an exponent:

$$\text{For } a \neq 0,\ a^0 = 1$$

So, for example, $(15456)^0 = 1$.

The key to negative exponents is that they always refer to reciprocals. So if $a \neq 0$ and n is a positive number,

$$a^{-n} = \frac{1}{a^n} \quad \text{and} \quad \frac{1}{a^{-n}} = a^n$$

7.7. NEGATIVE EXPONENTS

When simplifying things that have negative exponents, you don't want any negative exponents in your final answer.

Example: Simplify x^{-5}

Solution: $x^{-5} = \dfrac{1}{x^5}$

Example: Simplify $2x^{-5}$

Solution: $2x^{-5} = \dfrac{2}{x^5}$. (Notice that the 2 stayed on top because it didn't have a negative exponent. For comparison, see the next example.)

Example: Simplify $(2x)^{-5}$

Solution: In this expression, the exponent applies to the whole thing, so both the 2 and the x get moved to the bottom. Then we use our properties of exponents from Section 1.4 to finish up:

$$(2x)^{-5} = \dfrac{1}{(2x)^5} = \dfrac{1}{2^5 x^5} = \dfrac{1}{32x^5}$$

All of the properties of exponents we saw back in Section 1.4 also work for negative exponents, so we can simplify fairly messy looking expressions. As we go through the examples, keep in mind that there's usually more than one way to do the problem.

Example: Simplify $(2x^2 y^{-3})^2 (x^{-2} y^4)^{-3}$

Solution: Here's how I would do it (in great gory detail):

$$\begin{aligned}
(2x^2 y^{-3})^2 (x^{-2} y^4)^{-3} &= (2^2 (x^2)^2 (y^{-3})^2)((x^{-2})^{-3}(y^4)^{-3}) \\
&= (4x^{2 \cdot 2} y^{-3 \cdot 2}(x^{(-2) \cdot (-3)} y^{4 \cdot (-3)}) \\
&= (4x^4 y^{-6})(x^6 y^{-12}) \\
&= 4x^{4+6} y^{-6+(-12)} \\
&= 4x^{10} y^{-18} \\
&= \dfrac{4x^{10}}{y^{18}}
\end{aligned}$$

Example: Simplify $\dfrac{(2x^5y^{-7})^{-3}}{(3x^{-4}y^5)^2}$

Solution:

$$\begin{aligned}\dfrac{(2x^5y^{-7})^{-3}}{(3x^{-4}y^5)^2} &= \dfrac{2^{-3}x^{-15}y^{21}}{3^2x^{-8}y^{10}}\\ &= \dfrac{2^{-3}x^{-15-(-8)}y^{21-10}}{9}\\ &= \dfrac{2^{-3}x^{-7}y^{11}}{9}\\ &= \dfrac{y^{11}}{2^3\cdot 9x^7}\\ &= \dfrac{y^{11}}{8\cdot 9x^7}\\ &= \dfrac{y^{11}}{72x^7}\end{aligned}$$

Chapter 8

Radical Expressions

The next type of expression we'll look at is a **radical expression**, which is any expression that contains a radical (square root, cube root, etc.) These turn up whenever you need to undo some exponent.

8.1 ABCs of Radicals

What's Important

- Find roots of numbers.
- Simplify number radicals.

At this point, you should know all about raising things to a power. So, for instance,

$$5^2 = 25$$

Now we want to go backwards, meaning, we want to find a number whose square is 25. Obviously, the answer is 5, right? Actually, there's a second answer: -5. (Check it out, $(-5)^2 = 25$.) Both 5 and -5 are called **square roots** of 25 (5 is called the *principal* square root of 25). Note that you can only take square roots of positive numbers (or zero) and there are always two answers (except for zero), one positive, one negative. The notation is this:

$$\text{Principal square root of } a: \sqrt{a}$$

(The other square root is $-\sqrt{a}$.)

For some numbers, you can calculate the square root exactly:

Example: Find $\sqrt{49}$

Solution: Remember that $49 = 7^2$, so $\sqrt{49} = 7$.

For other numbers, you can only approximate the square root:

Example: Find $\sqrt{47}$

Solution: To approximate $\sqrt{47}$, we'll use a calculator. Most (scientific) calculators have a button for doing square roots so we'll use that. My calculator tells me

$$\sqrt{47} \approx 6.8556546$$

We don't have to stop with asking what number, when squared, gives (say) 25. We can ask, "What number, when cubed, gives 25?" or "What number, when raised to the fourth power, gives 25?" or.... Then we would be talking about cube roots, or fourth roots, or what have you. But the idea and the notation is still pretty much the same as for cube roots:

If $x^n = a$, then we say x is an nth root of a.

Principal nth root of a: $\sqrt[n]{a}$

(Some terminology: the thing under the radical sign is called the **radicand**. The little number in the v-shaped part of the radical sign is the **index**.)

Example: Find $\sqrt[3]{64}$

Solution: $64 = 4^3$, so $\sqrt[3]{64} = 4$.

Example: Find $\sqrt[7]{32}$

Solution: For this example, you need to use a calculator. Most scientific calculators have a root button or option, usually closely associated with the "raising to a power" button (read your manual). My calculator gives me this result:

$$\sqrt[7]{32} \approx 1.640670712$$

Some properties of roots (more later):

1. Ignoring zero, even roots always come in pairs (positive (principal root) and negative), while odd roots are loners. (Only one odd root, and it has the same sign as the radicand.)

8.1. ABCS OF RADICALS

2. If n is even, then $\sqrt[n]{a^n} = |a|$.

3. If n is odd, then $\sqrt[n]{a^n} = a$.

What's up with the absolute value in the second property? Look at this example:

Example: Find $\sqrt[4]{(-4)^4}$

Solution: Let's just work it out from the inside out:

$$\sqrt[4]{(-4)^4} = \sqrt[4]{256} = 4 (= |-4|)$$

What happened here? We raised -4 to the fourth power and got a positive 256, then took the principal fourth root (which must be positive) and got a positive 4, which is the same thing as the absolute value of -4.

So the reason that the second property has the absolute value in it is that all those even powers and principal even roots always give positive values.

We can use these properties to simplify some radical expressions.

Example: Simplify $\sqrt{25x^2}$

Solution: Rewrite $\sqrt{25x^2}$ as $\sqrt{(5x)^2}$ and use the second property:

$$\sqrt{(5x)^2} = |5x| = 5|x|$$

(Did you remember to use absolute value? We have an even root here.)

Example: Simplify $\sqrt[5]{32x^{10}}$

Solution: Rewrite $\sqrt[5]{32x^{10}}$ as $\sqrt[5]{(2x^2)^5}$. (Why x^2? Because $10 = 2 \cdot 5$, so $(x^2)^5 = x^{2 \cdot 5} = x^{10}$.) Then use the third property:

$$\sqrt[5]{(2x^2)^5} = 2x$$

By the way, most textbook writers eventually get bored with always throwing in the absolute value sign, so they start assuming that variables always stand for positive numbers. This allows them to leave it out. It's not wrong to keep it in there (where appropriate), but if you do, your answer may not match the one in the book. (I get bored, too, so I'm going to assume variables represent positive numbers from now on.)

8.2 Simplifying Radicals

What's Important

- Simplest radical form.
- Rationalizing a denominator.

Now for some more properties of radicals:

1. $\sqrt[n]{ab} = \sqrt[n]{a} \cdot \sqrt[n]{b}$

2. $\sqrt[n]{\dfrac{a}{b}} = \dfrac{\sqrt[n]{a}}{\sqrt[n]{b}}$

We'll use these properties to put expressions into what's usually called *simplest radical form*. The rules for this form are:

- Nothing under the radical has a power higher than the index.
- No fractions under the radical.
- No radicals in the denominator.

Example: Simplify $\sqrt{32}$

Solution: To simplify an expression like this, you want to use the properties above to pull out all squares (since we have a square root). So we'll start by writing this as square things times other stuff:

$$\sqrt{32} = \sqrt{16 \cdot 2} = \sqrt{4^2 \cdot 2}$$

Then we can use property 1 to break this up and simplify:

$$\sqrt{4^2 \cdot 2} = \sqrt{4^2} \cdot \sqrt{2} = 4\sqrt{2}$$

Notice that we used a property from Section 8.1 to change $\sqrt{4^2}$ to just 4.

Example: Simplify $\sqrt[3]{32x^3y^5}$

Solution: For this problem, we want to take out third powers, so we'll try to write it as third powers times other stuff:

$$\sqrt[3]{32x^3y^5} = \sqrt[3]{8 \cdot 4x^3y^3 \cdot y^2} = \sqrt[3]{2^3x^3y^3 \cdot 4y^2}$$

8.2. SIMPLIFYING RADICALS

Notice how we broke up the y^5 into $y^3 \cdot y^2$. (Remember that exponents *add*, so $y^5 = y^{3+2} = y^3 \cdot y^2$.) Now we'll use property 1 to break the radical up and simplify:

$$\sqrt[3]{2^3 x^3 y^3 \cdot 4y^2} = \sqrt[3]{2^3 x^3 y^3} \cdot \sqrt[3]{4y^2} = 2xy \sqrt[3]{4y^2}$$

Example: Simplify $\sqrt{\dfrac{8}{25}}$

Solution: Here we'll start by using property 2 from above:

$$\sqrt{\frac{8}{25}} = \frac{\sqrt{8}}{\sqrt{25}}$$

and then we simplify the top and the bottom:

$$\frac{\sqrt{8}}{\sqrt{25}} = \frac{\sqrt{4 \cdot 2}}{5} = \frac{\sqrt{4} \cdot \sqrt{2}}{5} = \frac{2\sqrt{2}}{5}$$

What about an expression such as $\dfrac{5}{\sqrt{2}}$? There's not much we can do to it to simplify the radical, but it's not in simplest radical form yet, since there's a $\sqrt{2}$ on the bottom. To deal with this situation, we need a technique called **rationalizing the denominator**. For square roots, this involves multiplying top and bottom by the radical on the bottom. For other roots, it's more complicated. You need to use the same kind of root, but you need to put things under the radical that complement what's already there so that you get everything up to the appropriate power. For instance, if you have $\sqrt[5]{x^2}$ on the bottom, you need to make that x have a fifth power, so you'll actually have to multiply by $\sqrt[5]{x^3}$. (I'll remind you again: Exponents add). Similarly, if you had $\sqrt[4]{8x^6 y^7} = \sqrt[4]{2^3 x^6 y^7}$ on the bottom, you need to make everything a fourth power (so all of the exponents are multiples of four), which means you would multiply by $\sqrt[4]{2x^2 y}$.

Example: Simplify $\dfrac{5}{\sqrt{2}}$

Solution: We have $\sqrt{2}$ on the bottom, so to rationalize the denominator we multiply top and bottom by $\sqrt{2}$:

$$\frac{5}{\sqrt{2}} = \frac{5 \cdot \sqrt{2}}{\sqrt{2} \cdot \sqrt{2}} = \frac{5\sqrt{2}}{2}$$

Example: Simplify $\dfrac{4x}{\sqrt[3]{4x^4y^2}}$

Solution: Here we have $\sqrt[3]{4x^4y^2} = \sqrt[3]{2^2x^4y^2}$ on the bottom, so we need to figure out what to multiply to make everything a cube. The answer is $\sqrt[3]{2x^2y}$. So multiply top and bottom by this and simplify.

$$\frac{4x}{\sqrt[3]{2^2x^4y^2}} \cdot \frac{\sqrt[3]{2x^2y}}{\sqrt[3]{2x^2y}} = \frac{4x\sqrt[3]{2x^2y}}{\sqrt[3]{2^3x^6y^3}} = \frac{4x\sqrt[3]{2x^2y}}{2x^2y} = \frac{2\sqrt[3]{2x^2y}}{xy}$$

(Remember to divide the 2 on the bottom into the 4 on top and the x on top into the x^2 on the bottom.)

8.3 Combining Radicals

What's Important

- Add, subtract, multiply and divide radicals.

Multiplying and dividing basic radicals is easy. In fact, it's just the properties from Section 8.2 written backward:

- $\sqrt[n]{a} \cdot \sqrt[n]{b} = \sqrt[n]{ab}$

- $\dfrac{\sqrt[n]{a}}{\sqrt[n]{b}} = \sqrt[n]{\dfrac{a}{b}}$

Example: Multiply: $\sqrt[3]{4} \cdot \sqrt[3]{4}$

Solution:
$$\sqrt[3]{4} \cdot \sqrt[3]{4} = \sqrt[3]{4 \cdot 4} = \sqrt[3]{16} = \sqrt[3]{8 \cdot 2} = \sqrt[3]{8} \cdot \sqrt[3]{2} = 2\sqrt[3]{2}$$

Example: Divide: $\dfrac{\sqrt{8}}{\sqrt{2}}$

Solution:
$$\frac{\sqrt{8}}{\sqrt{2}} = \sqrt{\frac{8}{2}} = \sqrt{4} = 2$$

8.3. COMBINING RADICALS

Adding and subtracting radicals is a little more complicated. It turns out that you can only add and subtract *like radicals*, by which I mean radicals that have the same index AND exactly the same thing under the radical. However, they can have different numerical coefficients. For example, $4\sqrt{7}$ and $-2\sqrt{7}$ are like radicals (both have $\sqrt{7}$) while $4\sqrt{7}$ and $4\sqrt[3]{7}$ are unlike radicals (different indexes), as are $4\sqrt{7}$ and $4\sqrt{5}$ (different radicands). Anyways, you add or subtract like radicals in the same way you add or subtract like terms in polynomials: Add or subtract the coefficients but keep the radical parts the same.

Example: Add: $4\sqrt{7} + 5\sqrt{7}$

Solution:
$$4\sqrt{7} + 5\sqrt{7} = (4+5)\sqrt{7} = 9\sqrt{7}$$

Example: Subtract: $4x\sqrt[3]{2xy} - 5x\sqrt[3]{2xy}$

Solution:
$$4x\sqrt[3]{2xy} - 5x\sqrt[3]{2xy} = (4x - 5x)\sqrt[3]{2xy} = -x\sqrt[3]{2xy}$$

Example: Add: $3\sqrt{50} + 5\sqrt{32}$

Solution: Uh-oh, unlike radicals! Actually, no. Let's start by simplifying the radicals:
$$3\sqrt{50} = 3\sqrt{25 \cdot 2} = 3\sqrt{25} \cdot \sqrt{2} = 3 \cdot 5 \cdot \sqrt{2} = 15\sqrt{2}$$

and
$$5\sqrt{32} = 5\sqrt{16 \cdot 2} = 5\sqrt{16} \cdot \sqrt{2} = 5 \cdot 4 \cdot \sqrt{2} = 20\sqrt{2}$$

So we have
$$3\sqrt{50} + 5\sqrt{32} = 15\sqrt{2} + 20\sqrt{2}$$

In other words, *like* radicals, so we can add them:
$$3\sqrt{50} + 5\sqrt{32} = 15\sqrt{2} + 20\sqrt{2} = 35\sqrt{2}$$

We may need to work with radical expressions that involve several different operations. Just follow the usual order of operations (PEMDAS). If you're multiplying radical binomials, think of FOILing out.

Example: Simplify $\sqrt{3}(\sqrt{2} + \sqrt{6})$

Solution: Start by distributing out:
$$\sqrt{3}(\sqrt{2} + \sqrt{6}) = \sqrt{3} \cdot \sqrt{2} + \sqrt{3} \cdot \sqrt{6}$$
then multiply and simplify:
$$\sqrt{6} + \sqrt{18} = \sqrt{6} + \sqrt{9 \cdot 2} = \sqrt{6} + 3\sqrt{2}$$

Example: Simplify $(\sqrt{2} + \sqrt{5})(\sqrt{2} + \sqrt{5})$

Solution: In this case, we'll FOIL it out and combine like terms:
$$\begin{aligned}(\sqrt{2} + \sqrt{5})(\sqrt{2} + \sqrt{5}) &= \sqrt{2} \cdot \sqrt{2} + \sqrt{2} \cdot \sqrt{5} + \sqrt{5} \cdot \sqrt{2} + \sqrt{5} \cdot \sqrt{5} \\ &= 2 + \sqrt{10} + \sqrt{10} + 5 \\ &= 7 + 2\sqrt{10}\end{aligned}$$

This last kind of problem shows up in another situation: rationalizing the denominator of things that look like (for example)
$$\frac{6}{\sqrt{2} + \sqrt{5}}$$
To do this, you multiply top and bottom by what's called the **conjugate** of the bottom, which is the same expression but with the opposite sign in the middle. So the conjugate of $\sqrt{2} + \sqrt{5}$ is $\sqrt{2} - \sqrt{5}$. If we multiply by it, here's what happens:
$$\begin{aligned}\frac{6}{\sqrt{2} + \sqrt{5}} \cdot \frac{\sqrt{2} - \sqrt{5}}{\sqrt{2} - \sqrt{5}} &= \frac{6(\sqrt{2} - \sqrt{5})}{(\sqrt{2} + \sqrt{5})(\sqrt{2} - \sqrt{5})} \\ &= \frac{6(\sqrt{2} - \sqrt{5})}{\sqrt{2} \cdot \sqrt{2} - \sqrt{2} \cdot \sqrt{5} + \sqrt{5} \cdot \sqrt{2} - \sqrt{5} \cdot \sqrt{5}} \\ &= \frac{6(\sqrt{2} - \sqrt{5})}{2 - \sqrt{10} + \sqrt{10} - 5} \\ &= \frac{6(\sqrt{2} - \sqrt{5})}{-3} \\ &= -2(\sqrt{2} - \sqrt{5}) \\ &= -2\sqrt{2} + 2\sqrt{5}\end{aligned}$$

8.4. SOLVING RADICAL EQUATIONS

So the radicals completely disappeared from the bottom! This will always happen when you multiply by the conjugate. The reason? Think of the formula for the difference of two squares.

8.4 Solving Radical Equations

What's Important

- Solve radical equations.

If an equation contains a radical of some kind, then it's called a **radical equation** (naturally). To solve such an equation, you always start by getting rid of the radical. The standard steps are:

1. Get the radical by itself on one side.

2. Raise both sides of the equation to the power that gets rid of the radical.

3. Repeat Steps 1 and 2 if necessary.

4. Solve the equation.

5. Check your answers in the original equation.

Note that the last step is to "check your answers in the original equation." This is because Step 2 can lead to values that aren't actually solutions. (Recall that when we solved rational equations, we also had this problem.)

Example: Solve $\sqrt{x+3} = 3$

Solution: The radical is already by itself, so we'll start by squaring both sides to get rid of the square root, then solve the equation:

$$(\sqrt{x+3})^2 = 3^2$$
$$x + 3 = 9$$
$$x = 6$$

And now we check our answer. Does the positive square root of $\sqrt{6+3} = \sqrt{9}$ equal 3? Yes it does, so the solution is $x = 6$.

Example: Solve $\sqrt{x+3} = x - 3$

Solution: Again, the radical is by itself, so we square both sides and solve:

$$(\sqrt{x+3})^2 = (x-3)^2$$
$$x + 3 = x^2 - 6x + 9$$
$$0 = x^2 - 7x + 6$$
$$0 = (x-6)(x-1)$$

Did you do the $(x-3)^2$ properly? Anyway, we get $x = 6$ or $x = 1$. We need to check both of these in the original equation:

$$\sqrt{6+3} = \sqrt{9} = 3 = 6 - 3$$

so $x = 6$ works.

$$\sqrt{1+3} = \sqrt{4} = 2 \neq 1 - 3$$

so $x = 1$ doesn't work. So the only solution is $x = 6$.

Example: Solve $\sqrt{x+3} - \sqrt{x-2} = 1$

Solution: For this problem, we have to start by getting one of the radicals by itself. Let's move the $\sqrt{x-2}$ over to the right side:

$$\sqrt{x+3} = 1 + \sqrt{x-2}$$

Then we square both sides. Be very careful in squaring the right-hand side!

$$(\sqrt{x+3})^2 = (1 + \sqrt{x-2})^2$$
$$x + 3 = (1 + \sqrt{x-2})(1 + \sqrt{x-2})$$
$$x + 3 = 1 + \sqrt{x-2} + \sqrt{x-2} + x - 2$$
$$x + 3 = x - 1 + 2\sqrt{x-2}$$

Notice that we still have a radical hanging around. So we have to get it by itself, and square

8.4. SOLVING RADICAL EQUATIONS

both sides again.

$$x + 3 = x - 1 + 2\sqrt{x-2}$$
$$4 = 2\sqrt{x-2}$$
$$4^2 = (2\sqrt{x-2})^2$$
$$16 = 4(x-2)$$
$$16 = 4x - 8$$
$$24 = 4x$$
$$6 = x$$

Did you remember to square the 2 that was in front of the $\sqrt{x-2}$? You may find it easier to divide by that 2 before you square both sides. Let's check $x = 6$ in the original equation:

$$\sqrt{6+3} - \sqrt{6-2} = \sqrt{9} - \sqrt{4} = 3 - 2 = 1$$

so $x = 6$ works.

For a last example, let's try a radical other than a square root.

Example: Solve $\sqrt[5]{x+3} = 2$

Solution: The radical is by itself, so we can immediately go to Step 2. To get rid of a fifth root, we raise both sides to the fifth power, then solve.

$$(\sqrt[5]{x+3})^5 = 2^5$$
$$x + 3 = 32$$
$$x = 29$$

As usual, check your answer in the original equation:

$$\sqrt[5]{29+3} = \sqrt[5]{32} = 2$$

So $x = 29$ works.

8.5 The Pythagorean Theorem

> **What's Important**
>
> - Do story problems using the Pythagorean Theorem.

The **Pythagorean Theorem** is one of the most famous statements in all of math. If you know just one theorem about right triangles, this is probably the theorem. Here's the version most students seem to know:

Pythagorean Theorem: If you have a right triangle labeled like this:

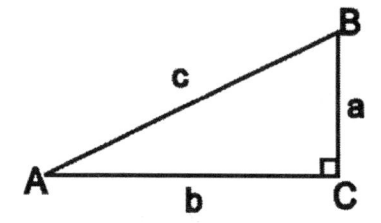

then $a^2 + b^2 = c^2$.

It's important to remember that c always means the side opposite the right angle (the *hypotenuse*), and that a and b are the other sides (the *legs*), since in problems that use this theorem, you may be finding any of the sides, so you need to put your numbers and variables in the right place.

Example: A 20-foot pole has a wire running from its top to a point 8 feet from its base. How long is the wire?

Solution: If you draw a picture, you'll see that you'll get a right triangle. The pole and the ground are the legs, and the wire is the hypotenuse. In other words, we have $a = 20$ feet, $b = 8$ feet and we want to find c. So apply the Pythagorean theorem:

$$a^2 + b^2 = c^2$$
$$20^2 + 8^2 = c^2$$
$$400 + 64 = c^2$$
$$464 = c^2$$

Then if we take square roots of both sides (use a calculator), we get $c = \pm 21.54$. A negative answer doesn't make sense as a length, so throw it out. So the wire is about 21.54 feet long.

Example: A pole has a 20-foot wire running from its top to a point 8 feet from its base. How tall is the pole?

Solution: Again, the picture looks like a right triangle with the pole and the ground as the legs and the wire as the hypotenuse. (You can re-use your picture for the last example if you relabel it.) This time, though, we have $b = 8$ feet, $c = 20$ feet and we want to find a. But we still use the Pythagorean Theorem:

$$a^2 + b^2 = c^2$$
$$a^2 + 8^2 = 20^2$$
$$a^2 + 64 = 400$$
$$a^2 = 400 - 64 = 336$$

Take square roots, and we get $a = \pm 18.33$. Again, throw out the negative value, so the solution is that the pole is about 18.33 feet tall.

8.6 Fractional Exponents

What's Important

- Evaluate expressions with fractional exponents.
- Simplify expressions with fractional exponents.

In Section 7.7 we extended what we know about exponents to negative exponents. In this section, we extend things to using fractions as exponents. The basic definition:

$$a^{1/n} = \sqrt[n]{a}$$

So, for example, $5^{1/3} = \sqrt[3]{5}$ and $16^{1/2} = \sqrt{16} = 4$.

You have to be careful with your minus signs when working with fractional exponents. For instance, $-16^{1/2}$ really means that you do the exponent first, then apply the minus sign, so $-16^{1/2} = -\sqrt{16} = -4$. On the other hand $(-16)^{1/2}$ means apply the minus sign first, then the exponent, so $(-16)^{1/2} = \sqrt{-16}$, not a real number.

We can extend our definition above to use exponents where the top is different from one.

$$a^{m/n} = \sqrt[n]{a^m} = (\sqrt[n]{a})^m$$

The important things to remember here:

- The numerator is the power.

- The denominator is the root.

Though it's possible to do the power and the root in either order, it's usually easier to do the root first when evaluating number expressions with fractional exponents.

Example: Simplify $16^{3/4}$

Solution: We'll do the root (bottom) first, then the power (top):

$$16^{3/4} = (\sqrt[4]{16})^3 = 2^3 = 8$$

Note that in the last example, we could have done the power first, then the root, but it would involve larger numbers:

$$16^{3/4} = \sqrt[4]{16^3} = \sqrt[4]{4096} = 8$$

All of the properties we've seen for exponents still work for fractional exponents. So you still add exponents when you multiply, you still subtract exponents when you divide, you still multiply exponents when you raise to a power and negative exponents still mean take the reciprocal.

Example: Simplify $(4x^{-3})^{3/2}(5x^{5/2})$

Solution: Apply the exponent in the first factor, and continue on with the usual rules:

$$\begin{aligned}(4x^{-3})^{3/2}(5x^{5/2}) &= (4^{3/2}(x^{-3})^{3/2})(5x^{5/2}) \\ &= (8x^{-9/2})(5x^{5/2}) \\ &= 40x^{-9/2+5/2} \\ &= 40x^{-4/2} = 40x^{-2} \\ &= \frac{40}{x^2}\end{aligned}$$

By switching between radical and exponential form, it's easier to simplify some expressions.

8.7. COMPLEX NUMBERS

Example: Simplify $\sqrt{x}\sqrt[3]{x^2}$

Solution: Start by changing to exponential form, then simplify and change back to radical form.
$$\sqrt{x}\sqrt[3]{x^2} = x^{1/2} \cdot x^{2/3} = x^{1/2+2/3} = x^{3/6+4/6} = x^{7/6} = \sqrt[6]{x^7} = x\sqrt[6]{x}$$

8.7 Complex Numbers

> **What's Important**
> - Add, subtract, multiply and divide complex numbers.

So far in this chapter, whenever we've tried to do the square root of a negative number, we've always had to say "You can't do this" or "Not a real number". And it's true, it's not a real number. Instead, it's what mathematicians call a **complex number**. To talk about complex numbers, we define a new symbol, i:

$$i = \sqrt{-1} \text{ (so } i^2 = -1)$$

First we use this for square roots of any negative number:

$$\sqrt{-a} = i\sqrt{a}$$

For example, $\sqrt{-16} = i\sqrt{16} = 4i$. Then we define a complex number to be something that looks like $a + bi$, where a and b are any real numbers. Some examples of complex numbers:

$$4 + 2i,\ 5 - 3i,\ -7i,\ 5$$

Notice that we consider real numbers to be complex numbers also. Some terminology:

- A number like $-7i$ is called an **imaginary number**.
- In a number like $4 + 2i$, the 4 is called the **real part** and the 2 is called the **imaginary part**.

Adding and subtracting and multiplying complex numbers is just like adding and subtracting binomials – just combine like terms (real parts with real parts, imaginary parts with imaginary parts).

Example: Add $(4+3i) + (5-2i)$

Solution:
$$(4+3i) + (5-2i) = (4+5) + (3i-2i) = 9+i$$

Example: Subtract $(4+3i) - (5-2i)$

Solution:
$$(4+3i) - (5-2i) = (4-5) + (3i-(-2i)) = -1+5i$$

To multiply complex numbers, FOIL them out and use the rule that $i^2 = -1$, then combine like terms.

Example: Multiply $(4+3i)(5-2i)$

Solution:
$$(4+3i)(5-2i) = 20 - 8i + 15i - 6i^2 = 20 - 8i + 15i - 6(-1) = 20 + 7i + 6 = 26 + 7i$$

Finally, to divide complex numbers, we need to use what's called the **complex conjugate**. Do you remember back in Section 8.3 how we rationalized some denominators by multiplying top and bottom by the conjugate of the bottom? What was the conjugate? It was the the same expression with the opposite sign in the middle. The complex conjugate works exactly the same way: just switch the sign in the middle from plus to minus or minus to plus. (To be more precise, you change the sign on the imaginary part.) So, for example, the complex conjugate of $5 + 2i$ is $5 - 2i$ and the complex conjugate of $4 - 7i$ is $4 + 7i$. For a trickier example, try the complex conjugate of $5i$. The answer is: $-5i$.

To actually divide two complex numbers, we multiply top and bottom by the complex conjugate of the one on the bottom and simplify.

Example: Divide $\dfrac{4+3i}{5-2i}$

Solution: Start by multiplying top and bottom by $5+2i$, which is the complex conjugate of the

8.7. COMPLEX NUMBERS

bottom. Then simplify.

$$\begin{aligned}\frac{4+3i}{5-2i} \cdot \frac{5+2i}{5+2i} &= \frac{20+8i+15i+6i^2}{25+10i-10i-4i^2} \\ &= \frac{20+23i+6(-1)}{25-4(-1)} \\ &= \frac{14+23i}{29} \\ &= \frac{14}{29} + \frac{23}{29}i\end{aligned}$$

Chapter 9

Quadratic Things

In this chapter, we take a closer look at quadratic equations and functions. We'll graph some functions (Section 9.1), look at some new ways to solve quadratic equations (Sections 9.2 and 9.3), solve some other equations (Section 9.4) and solve quadratic inequalities (Section 9.5).

9.1 Graphing a Quadratic Function

What's Important

- Graph a parabola by finding its vertex and intercepts.

A **quadratic function** is a function whose formula is a second degree polynomial. So a quadratic function looks like this:

$$f(x) = ax^2 + bx + c, \text{ where } a \neq 0$$

The graph of a quadratic function is a curve called a **parabola**. Here are two examples of parabolas, showing off the two basic types:

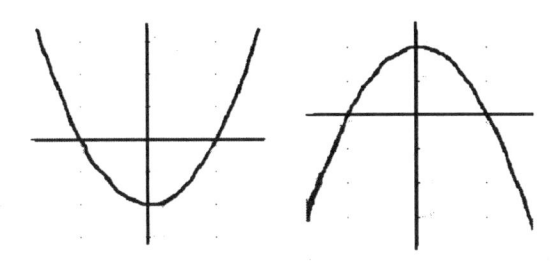

Notice that the parabola on the left has a lowest point, and the one on the right has a highest point. This lowest or highest point is called the **vertex** of the parabola. If the parabola has two x–intercepts (like these do), then the vertex will always be halfway between them, so you could just average the x–coordinates of the intercepts. But there is a handy formula (not given in Streeter, et al. until Section 10.1) for the vertex: The x–coordinate will be $-b/2a$ and the y–coordinate will be $f(-b/2a)$, where a is the coefficient of the x^2 in the function, and b is the coefficient of the x in the function. For example, the vertex of $f(x) = x^2 + 4x + 3$ will have x–coordinate $-4/(2)(1) = -2$ and y–coordinate $f(-2) = (-2)^2 + 4(-2) + 3 = 4 - 8 + 3 = -1$.

In this section, we're just going to work with quadratic functions that we can factor. The advantage of this is that we can use what we know about solving quadratic equations by factoring (see Section 6.9) to find the x–intercepts. Remember that x–intercepts are where the graph crosses the x-axis, so to find them algebraically, we set the function equal to zero and solve. So, for example, what are the x–intercepts of $f(x) = x^2 + 4x + 3$? Well, set it equal to zero and solve for x: $x^2 + 4x + 3 = 0$, so $(x+3)(x+1) = 0$, so $x = -3$ or $x = -1$. So the x–intercepts are $(-3, 0)$ and $(-1, 0)$.

We actually have enough information now to sketch a graph of the function. To do this, we plot the vertex and the x–intercepts and connect the dots with something that looks like a parabola:

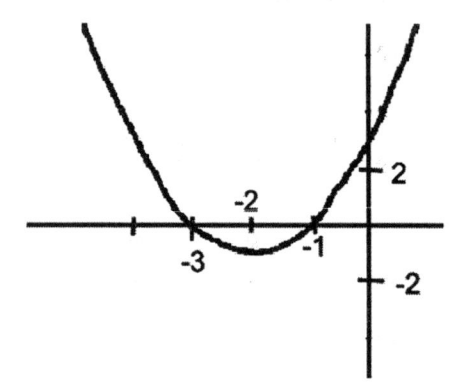

Let's try another example.

Example: Find the vertex and x–intercepts and sketch the graph of $f(x) = -x^2 + 2x$.

Solution: For the vertex, let's use the formula I gave above. The x–coordinate will be

$$\frac{-2}{2(-1)} = \frac{-2}{-2} = 1$$

The y–coordinate will be

$$f(1) = -1^2 + 2(1) = -1 + 2 = 1$$

So the vertex is $(1,1)$.

For the x-intercepts, set the function equal to zero and solve for x: $-x^2 + 2x = 0$, so $-x(x-2) = 0$, so $x = 0$ or $x = 2$. So the x-intercepts are $(0,0)$ and $(2,0)$.

Finally, to get the graph, plot the vertex and x-intercepts and connect the dots:

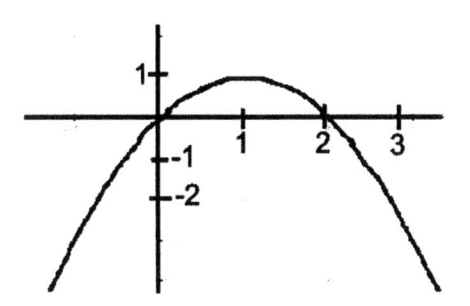

9.2 Completing the Square

What's Important
• Solve quadratic equations by completing the square.

We've looked at how to solve a quadratic equation by factoring. But not all quadratic equations can be factored. In this section and the next, we look at two ways to handle quadratic equations that we can't factor. As a first step, consider the equation $x^2 - 25 = 0$. This is the difference of squares, so we could factor it, but another way to do solve it is to move the 25 over to the right side and take positive and negative square roots:

$$x^2 - 25 = 0$$
$$x^2 = 25$$
$$x = \pm\sqrt{25} = \pm 5$$

This method works even when we don't have the difference of squares.

Example: Solve $x^2 - 24 = 0$

Solution: To solve this one, move the 24 over to the right side and take positive and negative square roots:

$$x^2 - 24 = 0$$
$$x^2 = 24$$
$$x = \pm\sqrt{24} = \pm 2\sqrt{6}$$

In fact, this method works for more complicated looking equations.

Example: Solve $(x+3)^2 = 14$

Solution: Notice that we still have something squared equalling a number, so we can take positive and negative square roots and solve for x:

$$(x+3)^2 = 14$$
$$x + 3 = \pm\sqrt{14}$$
$$x = -3 \pm \sqrt{14}$$

So we get two solutions, $x = -3 + \sqrt{14}$ and $x = -3 - \sqrt{14}$.

Now let's consider the equation $x^2 + 6x - 5 = 0$. Try factoring it. Can't do it, can you? So to solve it, we're going to use a technique called **completing the square**. The idea of this technique is to try to rewrite the equation so that it looks like something squared equals a number (just like the last example). Here are the steps:

1. Get the x's by themselves on the left side.

2. Divide everything by the coefficient of the x^2 if necessary.

3. Take half of the coefficient of the x and square it. Add this number to both sides of the equation.

4. The left side is now a perfect square, so you can factor it.

5. Take positive and negative square roots and solve for x.

Example: Solve $x^2 + 6x - 5 = 0$

Solution: Start by moving the 5 over to the right side:

$$x^2 + 6x = 5$$

9.2. COMPLETING THE SQUARE

We can skip step 2 here, so let's go straight to step 3. Half of 6 is 3, and 3 squared is 9. So we'll add 9 to both sides and factor the left side:

$$x^2 + 6x = 5$$
$$x^2 + 6x + 9 = 5 + 9$$
$$(x + 3)^2 = 14$$

And now if you look closely, you'll see we have the same equation as in the last example, so when we take square roots we again get $x = -3 + \sqrt{14}$ and $x = -3 - \sqrt{14}$.

Example: Solve $2x^2 + 4x - 10 = 0$

Solution: We'll start by moving the 10 to the right side, then we'll divide everything by 2:

$$2x^2 + 4x - 10 = 0$$
$$2x^2 + 4x = 10$$
$$x^2 + 2x = 5$$

Then to complete the square, we take half of 2, which is 1 and square it, which is still 1. So add 1 to both sides and factor the left side:

$$x^2 + 2x = 5$$
$$x^2 + 2x + 1 = 5 + 1$$
$$(x + 1)^2 = 6$$

So when we take square roots and solve for x, we get $x = -1 + \sqrt{6}$ and $x = -1 - \sqrt{6}$.

Example: Solve $x^2 - 4x + 6 = 0$

Solution: Start by moving the 6 to the right side:

$$x^2 - 4x = -6$$

Then take half of 4 and square it. So add 4 to both sides and factor the left side:

$$x^2 - 4x = -6$$
$$x^2 - 4x + 4 = -6 + 4$$
$$(x - 2)^2 = -2$$

In this case, when we take square roots we'll get complex solutions:

$$x = 2 + i\sqrt{2} \text{ and } x = 2 - i\sqrt{2}$$

9.3 The Quadratic Formula

What's Important

- Solve quadratic equations by using the quadratic formula.
- Use the discriminant to count the solutions of a quadratic equation.

The last method we'll look at for solving quadratic equations is the **quadratic formula**. Here it is: The solutions of $ax^2 + bx + c = 0$, where $a \neq 0$, are given by

$$x = \frac{-b \pm \sqrt{b^2 - 4ac}}{2a}$$

Learn this formula! If you take more math, you'll find you'll use it a lot. (By the way, this formula comes from completing the square on $ax^2 + bx + c = 0$. See most algebra textbooks for details.)

To actually use this formula, start by getting your equation in the right form (zero on one side), then figure out your a, b and c and plug them into the right places in the formula. To demonstrate, I'll repeat the last three examples from the last section, this time solving them using the quadratic formula.

Example: Solve $x^2 + 6x - 5 = 0$

Solution: We're already in the right form, so we can say $a = 1$, $b = 6$ and $c = -5$. So we plug

9.3. THE QUADRATIC FORMULA

these into the formula:

$$\frac{-b \pm \sqrt{b^2 - 4ac}}{2a} = \frac{-6 \pm \sqrt{6^2 - 4(1)(-5)}}{2(1)}$$
$$= \frac{-6 \pm \sqrt{36 - (-20)}}{2}$$
$$= \frac{-6 \pm \sqrt{56}}{2}$$
$$= \frac{-6 \pm 2\sqrt{14}}{2}$$
$$= -3 \pm \sqrt{14}$$

Example: Solve $2x^2 + 4x - 10 = 0$

Solution: For this equation $a = 2$, $b = 4$ and $c = -10$. So the formula gives us:

$$\frac{-b \pm \sqrt{b^2 - 4ac}}{2a} = \frac{-4 \pm \sqrt{4^2 - 4(2)(-10)}}{2(2)}$$
$$= \frac{-4 \pm \sqrt{16 - (-80)}}{4}$$
$$= \frac{-4 \pm \sqrt{96}}{4}$$
$$= \frac{-4 \pm 4\sqrt{6}}{4}$$
$$= -1 \pm \sqrt{6}$$

Example: Solve $x^2 - 4x + 6 = 0$

Solution: For this equation, $a = 1$, $b = -4$ and $c = 6$. Put these in the formula:

$$\frac{-b \pm \sqrt{b^2 - 4ac}}{2a} = \frac{-(-4) \pm \sqrt{(-4)^2 - 4(1)(6)}}{2(1)}$$
$$= \frac{4 \pm \sqrt{16 - 24}}{2}$$
$$= \frac{4 \pm \sqrt{-8}}{2}$$
$$= \frac{4 \pm (2\sqrt{2})i}{2}$$
$$= 2 \pm i\sqrt{2}$$

Notice that we got exactly the same solutions in all three examples.

So now we have three ways to solve quadratic equations; which is the best way? The answer is that it depends. If you can factor the equation, that's usually the quickest way to solve the problem. If you can't factor it, use the quadratic formula. The quadratic formula will always give you the solutions. On the whole, only use completing the square to solve an equation if you're specifically told to do so. But remember the method; it has other uses.

The thing under the radical in the quadratic formula (the $b^2 - 4ac$) has a special name. It's called the **discriminant**, and you can use it to learn how many real solutions your quadratic equation has without actually finding them. It all depends on the sign of the discriminant.

$b^2 - 4ac > 0$	2 real solutions
$b^2 - 4ac = 0$	1 real solution
$b^2 - 4ac < 0$	no real solutions (2 nonreal solutions)

Example: How many real solutions does $x^2 - 4x + 4 = 0$ have?

Solution: To answer this question, use the discriminant with $a = 1$, $b = -4$ and $c = 4$. Then

$$b^2 - 4ac = (-4)^2 - 4(1)(4) = 16 - 16 = 0$$

so there is exactly one real solution.

9.4 More Equations

> **What's Important**
>
> - Solving equations that are quadratic in form.

In this section, we'll solve some equations that aren't quadratic, but do fit the pattern in a way. In other words, they look like a number times something squared plus a number times that something plus a number equals zero. Here's an example:

$$x^4 - 6x^2 + 5 = 0$$

Notice that if we write it like this:

$$(x^2)^2 - 6x^2 + 5 = 0$$

it fits the pattern of a number times something squared plus a number times that something plus a number equals zero.

To solve an equation like this, we're going to change variables to make it into a quadratic equation, solve that new equation, then change variables back and solve the leftover equations. A few examples are probably in order at this point.

Example: Solve $x^4 - 6x^2 + 5 = 0$

Solution: As mentioned above, this equation is quadratic in form like this:

$$(x^2)^2 - 6x^2 + 5 = 0$$

So we're going to change variables by letting $u = x^2$ and substituting this in:

$$u^2 - 6u + 5 = 0$$

Now we have an ordinary quadratic equation, so let's go ahead and solve it:

$$u^2 - 6u + 5 = 0$$
$$(u - 5)(u - 1) = 0$$

So $u = 5$ or $u = 1$. Now we have to change back to x, so we have $x^2 = 5$ or $x^2 = 1$. If we solve these by taking square roots we get $x = \pm\sqrt{5}$ or $x = \pm 1$. So we have four solutions. Note that you may want to check your answers in the original equation, since it is possible to get "bad" solutions.

Example: Solve $x - 5\sqrt{x} - 6 = 0$

Solution: If you remember Section 8.4, it's possible to solve this by getting the radical by itself and squaring both sides. But here we want to look at it as quadratic in form. So how do we do that? If you think of x as being the square of $\sqrt{x}$, then you can see this equation can be written as

$$(\sqrt{x})^2 - 5\sqrt{x} - 6 = 0$$

so it fits the pattern. So let's change variables by letting $u = \sqrt{x}$. When we substitute in we get

$$u^2 - 5u - 6 = 0$$

Now solve this in the usual way:

$$u^2 - 5u - 6 = 0$$
$$(u - 6)(u + 1) = 0$$

So $u = 6$ or $u = 1$. Now change back to x: $\sqrt{x} = 6$ or $\sqrt{x} = 1$. To solve these, square both sides to get $x = 36$ or $x = 1$. Remember to check your answers in the original equation! It turns out that $x = 36$ works, but $x = 1$ doesn't work, so this equation has only one solution: $x = 36$.

Example: Solve $(x^2 + 2x)^2 - 2(x^2 + 2x) - 3 = 0$

Solution: Though complicated looking, this equation is still quadratic in form. We'll let $u = x^2 + 2x$ and substitute this in to get

$$u^2 - 2u - 3 = 0$$

Factor and solve:

$$u^2 - 2u - 3 = 0$$
$$(u - 3)(u + 1) = 0$$

So $u = 3$ or $u = -1$. Change back to x and we have two new quadratic equations to solve:

$x^2 + 2x = 3$ and $x^2 + 2x = -1$. So solve the first one:

$$x^2 + 2x = 3$$
$$x^2 + 2x - 3 = 0$$
$$(x+3)(x-1) = 0$$

which gives us $x = -3$ or $x = 1$, and solve the second one:

$$x^2 + 2x = -1$$
$$x^2 + 2x + 1 = 0$$
$$(x+1)(x+1) = 0$$

which gives us $x = -1$. So in this case, we have three possible solutions $x = -3$, $x = 1$ and $x = -1$ and if we check them in the original equation, they all work.

9.5 Quadratic Inequalities

What's Important

- Solve quadratic inequalities algebraically.

A **quadratic inequality** is an inequality with an x^2 in it, but no higher power. Here are some examples:

$$x^2 - 4 > 0,\ x^2 + 3x \leq 4,\ 2x^2 + 5x + 2 > 0$$

Streeter, et al talk about solving these both graphically and algebraically. The graphical method is similar to what we looked at in Section 4.4 for linear inequalities – graph the function part and shade in a region or a segment. The only difference now is that you would graph parabolas rather than lines. Being able to solve them algebraically is much more important, so that's what I'll talk about. Here's the method:

1. Get zero on one side of the inequality.

2. Find where the quadratic part equals zero and plot those points on a number line.

3. Test values in between the zeros to determine whether the quadratic part is positive or negative on a particular interval.

4. For your answer, choose the intervals where the sign matches the inequality.

And here's an example:

Example: Solve $x^2 - 4 > 0$

Solution: We already have zero on one side, so we'll start by finding where $x^2 - 4$ equals zero.

$$x^2 - 4 = 0$$
$$(x-2)(x+2) = 0$$

So $x = 2$ or $x = -2$. Now let's put these on a number line:

We're going to test values other than ± 2 to see whether $x^2 - 4$ is positive or negative. Notice our number line got divided into three parts, and we need to check the sign on each part. So pick a number less than -2 (such as -3), a number between -2 and 2 (such as 0) and a number greater than 2 (such as 3) and plug them into $x^2 - 4$. Here are the results:

$x = -3$	$(-3)^2 - 4 = 9 - 4 = 5$	positive
$x = 0$	$0^2 - 4 = 0 - 4 = -4$	negative
$x = 3$	$3^2 - 4 = 9 - 4 = 5$	positive

So what's the solution? We're trying to solve $x^2 - 4 > 0$, so we want $x^2 - 4$ to be positive. So pick the parts of the number line where it is positive and we get the solution $\{x | x < -2 \text{ or } x > 2\}$ as a set, or as a graph

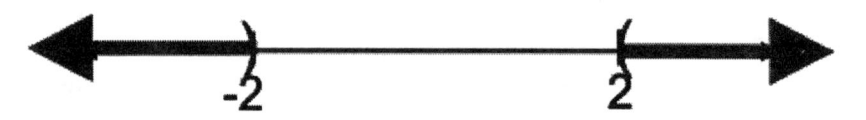

Example: Solve $x^2 + 3x \leq 4$

Solution: Start by getting zero on one side:

$$x^2 + 3x - 4 \leq 0$$

9.5. QUADRATIC INEQUALITIES

Then find where $x^2 + 3x - 4$ equals zero:

$$x^2 + 3x - 4 = 0$$
$$(x+4)(x-1) = 0$$

So $x = -4$ or $x = 1$. Let's put these on a number line:

Again, the number line is divided into three parts and we need to check whether $x^2 + 3x - 4$ is positive or negative on each part. So choose a number less than -4 (such as -5), a number between -4 and 1 (such as 0) and a number greater than 1 (such as 2). Here are the results:

$x = -5$	$(-5)^2 + 3(-5) - 4 = 25 - 15 - 4 = 6$	positive
$x = 0$	$0^2 + 3(0) - 4 = 0 + 0 - 4 = -4$	negative
$x = 2$	$2^2 + 3(2) - 4 = 4 + 6 - 4 = 6$	positive

We want $x^2 + 3x - 4$ to be less than or equal to zero, so we'll choose the part where it's negative and include the endpoints. So our solution set is $\{x| -4 \leq x \leq 1\}$ and the graph is

Chapter 10

Conic Sections

In this chapter we look at the family of curves called **conic sections**, so called because if you slice a cone with a plane, the cross–section will look like one of those curves. (Try this the next time you get ice cream.) The three kinds of conic sections are the *parabola*, the *ellipse* and the *hyperbola*.

10.1 Graphing Parabolas

What's Important

- Graph parabolas by shifting.

We've actually looked a little at parabolas back in Section 9.1, so some of what I say here will repeat what I said then, but some stuff will be new.

First off, the basic parabolas you should learn are $y = x^2$ and $y = -x^2$. Their graphs look like this:

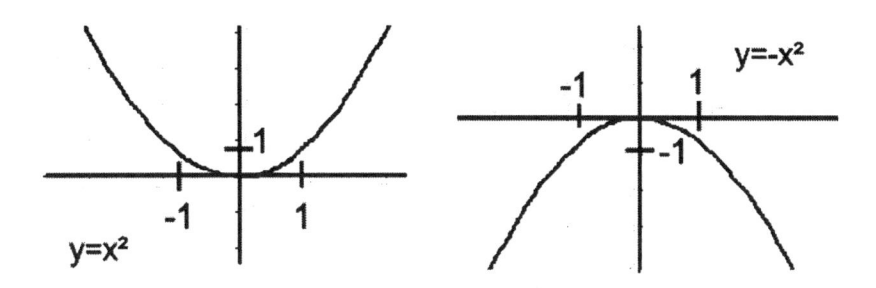

Both have their vertex at the origin. $y = x^2$ opens up, $y = -x^2$ opens down. (That minus sign in front actually means to flip over the x-axis, so they're actually mirror images of each other.) They have one more feature – an **axis of symmetry**. For these graphs, if you look at the y-axis, it runs right down the middle through the vertex, and the left half of the parabola is the mirror image of the right half. In terms of points, it means that for a particular point on the left side of the parabola, there is a corresponding point on the right side of the curve with opposite x–coordinate and the same y–coordinate. For example, the point $(-1, 1)$ is on the left side of $y = x^2$, and on the right side you can find the corresponding point $(1, 1)$. It turns out that all parabolas have an axis of symmetry running down their middle, and you can always find pairs of symmetric points (points the same distance on either side of the axis of symmetry). In fact, we'll use such points to help us sketch the graph. But first, here's the standard equation for a parabola that opens up or down:

$$y = a(x - h)^2 + k$$

This equation can tell you a lot about the parabola.

- The vertex is (h, k).

- The axis of symmetry is the vertical line $x = h$.

- If $a > 0$, the parabola opens up. If $a < 0$, the parabola opens down.

- The value of a tells you something about how wide or narrow the parabola is.

- The graph looks the same as the graph of $y = ax^2$, but shifted h units to the left (if you see $+h$) or right (if you see $-h$) and k units up (if you see $+k$) or down (if you see $-k$).

Example: Sketch the graph of $y = (x - 1)^2 - 2$

Solution: According to what I said above, the vertex of this parabola is $(1, -2)$, it has axis of symmetry $x = 1$ and it opens upward. To graph it, we start with $y = x^2$ and shift it 1 unit to the right and 2 units down. So the graph looks like the picture below. Notice that we have symmetric points at (for example) $(0, -1)$ and $(2, -1)$, both of which are 1 unit on either side of the axis of symmetry.

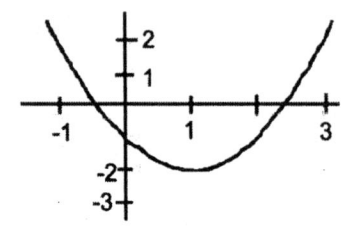

10.1. GRAPHING PARABOLAS

Of course, it's too much to expect that you'll be always handed the parabola equation in the right form. But you can always convert a quadratic function to the form above by completing the square.

Example: Sketch the graph of $y = x^2 - 2x - 3$

Solution: We'll start by putting it in the right form. To do this, move the 3 over to the left side and complete the square, then get the y by itself:

$$y = x^2 - 2x - 3$$
$$y + 3 = x^2 - 2x$$
$$y + 3 + 1 = x^2 - 2x + 1$$
$$y + 4 = (x - 1)^2$$
$$y = (x - 1)^2 - 4$$

So in standard form, our equation is $y = (x - 1)^2 - 4$. So the vertex is $(1, -4)$, the axis of symmetry is $x = 1$ and it opens upward. To graph it we can start with $y = x^2$ and shift it one unit to the right and four units down. Another way to do this is to find a pair of symmetric points. Just pick an x value on one side of the vertex (such as $x = 0$) and find the point there (in this case, $(0, -3)$. Then find the symmetric point, which is the same distance away from the axis of symmetry on the other side (for $(0, -3)$, the symmetric point would be $(2, -3)$). Then plot the vertex and the symmetric points and connect the dots to get your graph:

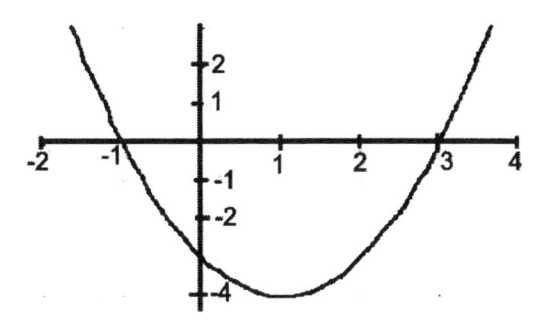

By the way, graphing by finding symmetric points is particularly useful when you have something other than 1 or -1 in front of the x^2, because then simply moving $y = x^2$ or $y = -x^2$ around doesn't work.

As a last note on parabolas, it's also possible to have parabolas that open to the left or to the right. The standard equation for these looks like

$$x = a(y - k)^2 + h$$

The vertex will still be at (h, k), the axis of symmetry will be the horizontal line $y = k$, it will open to the left if $a < 0$, to the right if $a > 0$ and you can still talk about symmetric points and moving graphs around ($x = y^2$ or $x = -y^2$ now). And you still use completing the square to get it in the right form.

Example: Graph $x = -y^2 + 2x + 2$

Solution: Start by moving the 2 to the left side, then complete the square and get the x by itself:

$$x = -y^2 + 2y + 2$$
$$x - 2 = -y^2 + 2y$$
$$-x + 2 = y^2 - 2y \text{ (We divided by } -1 \text{ to make life easier.)}$$
$$-x + 2 + 1 = y^2 - 2y + 1$$
$$-x + 3 = (y - 1)^2$$
$$x = -(y - 1)^2 + 3$$

So the equation in standard form is $x = -(y-1)^2 + 3$. The vertex is $(3, 1)$, the axis of symmetry is $y = 1$, it opens to the left and one pair of symmetric points is $(2, 0)$ and $(2, 2)$. So the graph looks like this:

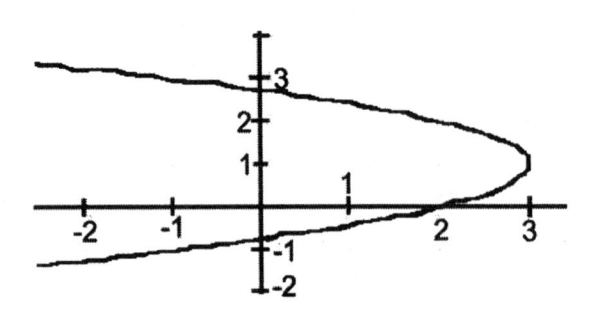

10.2 Circles

> **What's Important**
> - Find an equation of circle from its center and radius.
> - Find the center and radius of a circle from its equation.

Circles are one of the most easily recognized geometric figures. The standard equation for a circle that has its center at the point (h, k) and radius r is

$$(x - h)^2 + (y - k)^2 = r^2$$

(This formula, by the way, comes from the distance formula

$$d = \sqrt{(x_2 - x_1)^2 + (y_2 - y_1)^2}$$

which gives the distance between the points (x_1, y_1) and (x_2, y_2).)

Example: Find the equation of a circle with center $(4, -2)$ and radius 5.

Solution: Take the standard equation for a circle and put 4 in for h, -2 in for k and 5 in for r and you get

$$(x - 4)^2 + (y - (-2))^2 = 5^2$$

or

$$(x - 4)^2 + (y + 2)^2 = 25$$

Example: What is the center and radius of $(x+5)^2 + (y-3)^2 = 14$?

Solution: Let's rewrite the equation slightly:

$$(x-(-5))^2 + (y-3)^2 = (\sqrt{14})^2$$

Then if we compare this to the standard equation, we see that $h = -5$, $k = 3$ and $r = \sqrt{14}$, so the center is $(-5, 3)$ and the radius is $\sqrt{14}$.

Example: What is the center and radius of $x^2 + y^2 + 4x - 6y + 2 = 0$?

Solution: Though it may not look like it, this really is the equation of a circle. To answer the question, we have to put it in standard form. How do we do this? By completing the square. So move the 2 to the right side, group the x's and y's together and complete the square on both:

$$x^2 + y^2 + 4x - 6y + 2 = 0$$
$$x^2 + 4x + y^2 - 6y = -2$$
$$(x^2 + 4x + 4) + (y^2 - 6y + 9) = -2 + 4 + 9$$
$$(x+2)^2 + (y-3)^2 = 11$$
$$(x-(-2))^2 + (y-3)^2 = (\sqrt{11})^2$$

So if we compare this to the standard equation above, we see that $h = -2$, $k = 3$ and $r = \sqrt{11}$, so the center is $(-2, 3)$ and the radius is $\sqrt{11}$.

10.3 Ellipses and Hyperbolas

What's Important

- Graph an ellipse.
- Graph a hyperbola.

A circle is actually a special case of another kind of curve called an **ellipse**. See the top of the next page for an ellipse.

This happens to be the ellipse $\dfrac{x^2}{9} + \dfrac{y^2}{4} = 1$. This particular ellipse has its center at the origin and its longer direction (called the **major axis**) oriented horizontally. (The shorter direction,

10.3. ELLIPSES AND HYPERBOLAS

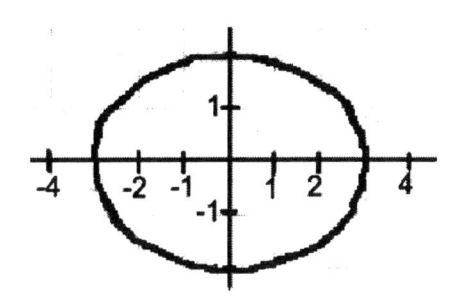

by the way, is called the **minor axis**.) The points at the ends of the major axis are called the **vertices** of the ellipse, so for this ellipse, the vertices are $(-3, 0)$ and $(3, 0)$. Notice that the x–coordinates are the positive and negative square roots of the number under the x^2 in the equation. It's also possible to have ellipses whose major axis is vertical.

We're just going to worry about ellipses that are centered at the origin. The standard equation for such an ellipse is

$$\frac{x^2}{a^2} + \frac{y^2}{b^2} = 1$$

Some comments on this equation:

- If $a > b$, then the major axis is horizontal and the vertices are $(\pm a, 0)$. The ends of the minor axis are $(0, \pm b)$.

- If $a < b$, then the major axis is vertical and the vertices are $(0, \pm b)$. The ends of the minor axis are $(\pm a, 0)$.

- If $a = b$, then we actually have a circle.

To graph an ellipse from its equation, plot the points at the end of the major and minor axes, then connect the dots with something that looks like an ellipse.

Example: Graph $16x^2 + 4y^2 = 16$

Solution: This is an equation of an ellipse, but it's not in standard form. To put it in standard form, just divide everything by 16 and simplify:

$$\frac{16x^2}{16} + \frac{4y^2}{16} = \frac{16}{16}$$
$$\frac{x^2}{1} + \frac{y^2}{4} = 1$$

If we compare this to the standard form, we see that $a = 1$ and $b = 2$. Since $a < b$, the major axis is vertical and the vertices are $(0, \pm 2)$ while the ends of the minor axis are $(\pm 1, 0)$. So plot these points and connect the dots to get your ellipse:

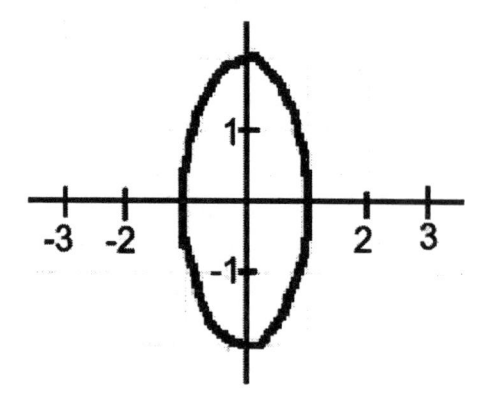

The last kind of conic section to consider is the **hyperbola**. The standard equation for a hyperbola (center at the origin) looks like this

$$\frac{x^2}{a^2} - \frac{y^2}{b^2} = 1$$

for a hyperbola that opens to the left and right, or

$$\frac{y^2}{b^2} - \frac{x^2}{a^2} = 1$$

for a hyperbola that opens up and down. Here's a picture of the first type. Its equation is $\frac{x^2}{4} - \frac{y^2}{9} = 1$.

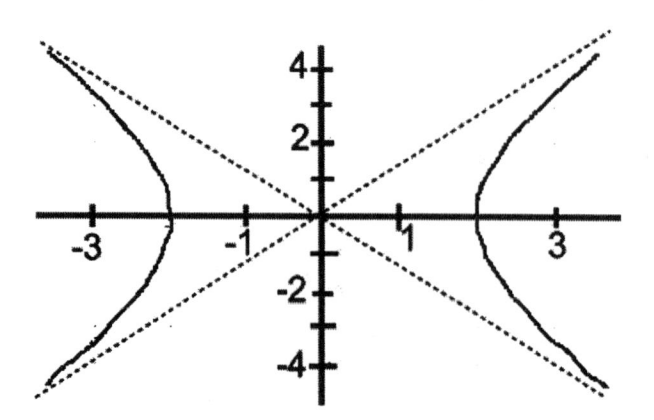

10.3. ELLIPSES AND HYPERBOLAS

The dotted lines that you see are not actually part of the hyperbola. Instead, those are lines (called **asymptotes**) that the parts of the hyperbola approach more and more closely the farther you go out from the origin. They're a great help in drawing the hyperbola properly. The equations of the asymptotes will always be

$$y = \pm \frac{b}{a} x$$

for the hyperbolas we're looking at (center at origin, equation as above).

Just like ellipses, hyperbolas have points called vertices. On a hyperbola these are the middle points of the branches, so their coordinates will be $(\pm a, 0)$ for a hyperbola that opens left and right or $(0, \pm b)$ for a hyperbola that opens up and down.

Graphing a hyperbola from its equation is a matter of plotting its vertices and asymptotes, then sketching something that looks like a hyperbola. Put your equation in standard form first, if necessary.

Example: Sketch the graph of $4y^2 - 16x^2 = 16$

Solution: Start by dividing everything by 16 to put it in standard form:

$$\frac{y^2}{4} - \frac{x^2}{1} = 1$$

Then if you compare it to the standard equations, you'll see that this hyperbola opens up and down with $b = 2$ and $a = 1$, so the vertices are $(0, \pm 2)$ and the asymptotes are $y = \pm 2x$. Plot these and draw your curve and you should end up with something like this:

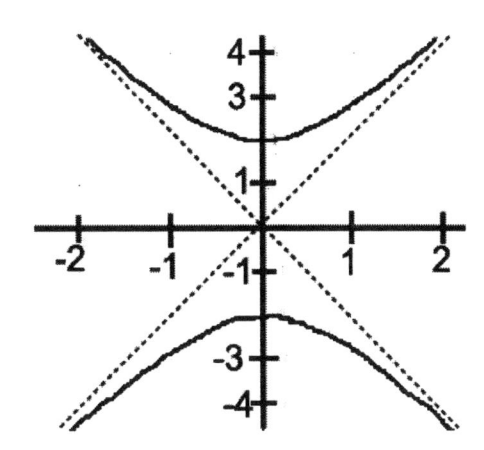

10.4 Nonlinear Systems

What's Important

- Solve a nonlinear system of equations.
- Graph a nonlinear system of inequalities.

We looked at solving systems of linear equations back in Chapter 5. In this section we look at **nonlinear systems** – systems of equations where at least one equation is not a line. The same methods we used for solving linear systems (graphing, substitution, elimination) will work for nonlinear systems, though, as before, the algebraic methods are better. But the geometric meaning is still looking where two curves intersect. Actually, substitution is probably the most commonly used method for solving nonlinear systems.

Example: Solve the system

$$3x + y = 4$$
$$y = x^2$$

Solution: Geometrically, in this system we're seeing where a line and a parabola intersect. As mentioned above, we'll try to solve this system by substitution. Notice that the second equation is already solved for y, so we'll take the x^2 and substitute it in for y in the first equation:

$$3x + x^2 = 4$$

Now move the 4 over to the left and we have a quadratic equation to solve:

$$3x + x^2 = 4$$
$$x^2 + 3x - 4 = 0$$
$$(x+4)(x-1) = 0$$

So we get $x = -4$ or $x = 1$. The corresponding y values are $y = 16$ and $y = 1$, respectively, so the two solutions are $(-4, 16)$ and $(1, 1)$.

10.4. NONLINEAR SYSTEMS

Example: Solve the system

$$x^2 + 4y^2 = 16$$
$$x^2 - y^2 = 1$$

Example: In this system, we have an ellipse and a hyperbola. It's actually easiest to solve this system by elimination. Multiply the second equation by 4 and add them together and you get

$$5x^2 = 20$$

If we solve this for x by dividing by 5 and taking positive and negative square roots we get

$$5x^2 = 20$$
$$x^2 = 4$$
$$x = \pm 2$$

Then to get the corresponding y values, plug these back into one of the original equations (the first, say).

$$\begin{array}{ll} x = 2 & x = -2 \\ (2)^2 + 4y^2 = 16 & (-2)^2 + 4y^2 = 16 \\ 4 + 4y^2 = 16 & 4 + 4y^2 = 16 \\ 4y^2 = 12 & 4y^2 = 12 \\ y^2 = 3 & y^2 = 3 \\ y = \pm\sqrt{3} & y = \pm\sqrt{3} \end{array}$$

So we end up with four solutions: $(2, \sqrt{3})$, $(2, -\sqrt{3})$, $(-2, \sqrt{3})$ and $(-2, -\sqrt{3})$.

We can also have systems of nonlinear inequalities. Back in Section 5.5, we solved systems of linear inequalities by graphing the lines and shading the appropriate region. A system of nonlinear inequalities is solved much the same way, by graphing the curves and shading the appropriate region.

Example: Graph the system

$$x + y \leq 1$$
$$y \geq x^2$$

Solution: Start by graphing the line $x + y = 1$ and the parabola $y = x^2$ on the same set of axes:

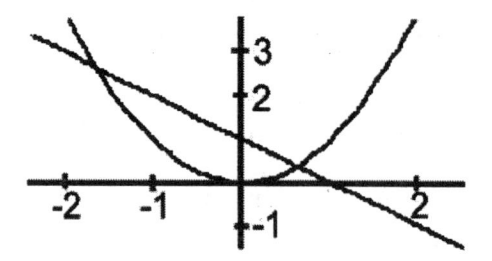

Then, according to our inequalities, we want to shade below the line $x + y = 1$ and above the parabola $y = x^2$, so we end up shading that little region between them:

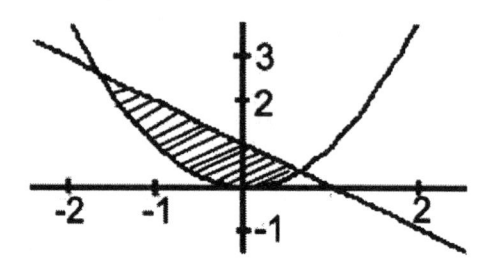

Chapter 11

Exponentials and Logarithms

The main topic of this chapter are two useful families of functions: exponential functions and logarithmic functions. These families turn out to be related to each other in a special way, namely, they're *inverses* of each other. Section 11.1 explains what we mean by that.

11.1 Inverse Functions

What's Important

- Find the inverse of a function.

Suppose you have a function, such as $f(x) = 4x + 3$. What is $f(2)$? Just put 2 in for x and you get
$$f(2) = 4(2) + 3 = 8 + 3 = 11$$
Now for a different question: Where does $f(x) = 5$? In this case, we have a function value and want to know what x value corresponds to it. That's what **inverse functions** are all about – running functions backwards, or starting with function values and finding the corresponding x value.

So how do we find an inverse function? It all depends on how the function is represented. If you think back to Chapter 3, we looked at several different ways to represent functions, for instance, as sets of ordered pairs, as formulas, and as graphs. For these, here's how to find the inverse:

- Sets of ordered pairs – Switch the x and y coordinates in the ordered pairs.

- Formulas – Switch x and y and solve for y.

- Graphs – Reflect the graph over the line $y = x$.

By the way, here's some notation. The inverse of f is written f^{-1}.

Example: Find the inverse of $\{(1,4),(3,2),(7,1)\}$.

Solution: We have a set of ordered pairs, so we just switch the coordinates around. So the inverse will be the set
$$\{(4,1),(2,3),(1,7)\}$$

Example: Find the inverse of $f(x) = 4x + 3$

Solution: We have a formula in this case, so we find the inverse by switching x and y (start by replacing $f(x)$ with y) and solving for y:

$$y = 4x + 3$$
$$x = 4y + 3$$
$$x - 3 = 4y$$
$$\frac{x-3}{4} = y$$

So $f^{-1}(x) = \dfrac{x-3}{4}$. (By the way, we can now answer the question about where does $f(x) = 5$? We just find $f^{-1}(5) = \dfrac{5-3}{4} = \dfrac{2}{4} = \dfrac{1}{2}$.

Example: Find the inverse of

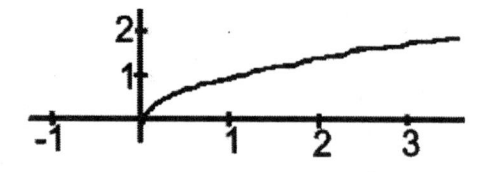

Solution: To find the inverse of this function, reflect the graph over the line $y = x$. You end up with a picture like this (the dotted curve is the original function, the solid curve is the inverse function):

You should be aware that not all functions have inverse functions. In order to have an inverse, a function must be **one-to-one**. This means that different x values always get matched to different function values. The standard example of a function that's not one-to-one is $f(x) = x^2$.

11.2. EXPONENTIALS

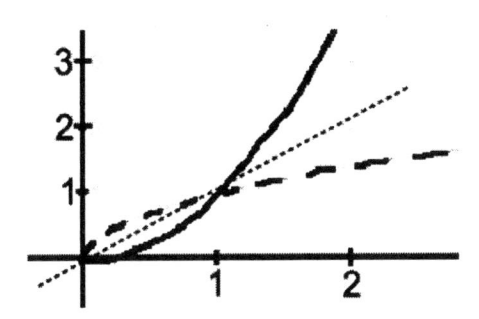

Note that $f(-2) = 4$ and $f(2) = 4$, so we have two different x values (2 and -2) that get matched to the *same* function value (4). That's what makes it not one-to-one.

In most cases, the easiest way to decide whether a function is one-to-one is the *horizontal line test*. Graph the function and draw horizontal lines. If you can draw a horizontal line that intersects the function more than once, then it's not one-to-one. If you can't draw such a line, then it is one-to-one. Let's try this with the graph of $f(x) = x^2$:

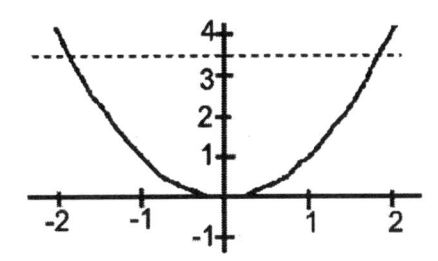

We were able to find a horizontal line that intersects it more than once, so it's not one-to-one.

Finally, the exact relation between inverse functions and one-to-one functions is this: Every function with an inverse is one-to-one and every one-to-one function has an inverse.

11.2 Exponentials

What's Important

- Graph exponential functions.
- Solve simple exponential equations.

Exponential functions are functions where the exponent is a variable. In symbols, an exponential function looks like $f(x) = b^x$, where b (the **base**) is positive and not equal to 1.

Here are some exponential functions:

$$f(x) = 2^x, \; f(x) = (1.457)^x, \; f(x) = \left(\frac{1}{5}\right)^x$$

As with all functions, to evaluate an exponential function, put the value in for the x and simplify.

Example: Let $f(x) = 2^x$. Find $f(4)$, $f(0)$ and $f(-2)$.

Solution:

$$f(4) = 2^4 = 16$$
$$f(0) = 2^0 = 1$$
$$f(-2) = 2^{-2} = \frac{1}{2^2} = \frac{1}{4}$$

As you can see, we use a number of the properties of exponents in evaluating an exponential function (as you might expect).

Graphing an exponential function is best done by making a table of values and plotting points. But you should become familiar with the two basic shapes for an exponential curve, as shown in the next two examples.

Example: Graph $f(x) = 2^x$

Solution: Make a table of values:

x	-2	-1	0	1	2
$f(x)$	$1/4$	$1/2$	1	2	4

Then plot these points and connect the dots:

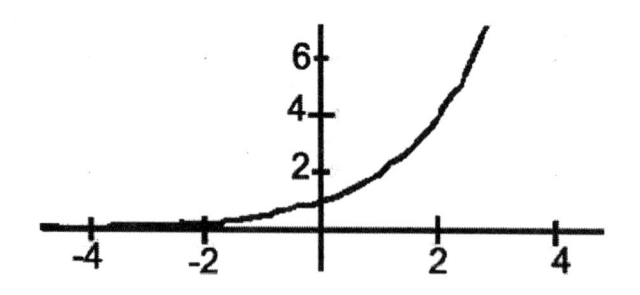

(When people talk about something increasing exponentially, think of a curve like this. It starts out slow, then grows faster and faster.)

11.2. EXPONENTIALS

Example: Graph $f(x) = \left(\dfrac{1}{2}\right)^x$

Solution: Again, make a table of values, then plot points and connect the dots.

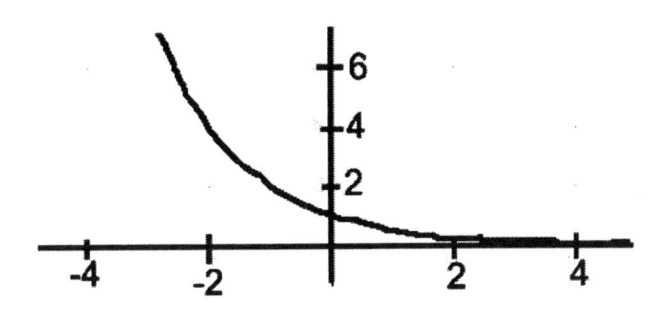

x	-2	-1	0	1	2
$f(x)$	4	2	1	$1/2$	$1/4$

Before taking a look at some simple exponential equations, there's one particular exponential function I want to mention, called the **natural exponential function**. It's written $f(x) = e^x$, where e is a particular irrational number. (To nine decimal places, $e \approx 2.718281828$.) This function appears in all kinds of situations, including population growth, radioactive decay, and continuously compounded interest.

Like every other kind of expression that we've looked at, exponentials can appear in equations. The simplest kind of exponential equations are solved using this property:

$$\text{If } b^m = b^n, \text{ then } m = n.$$

Example: Solve $3^x = 9$

Solution: Start by writing 9 in exponential form: $9 = 3^2$. So we actually have $3^x = 3^2$, so we can use the property above to say that $x = 2$.

Example: Solve $3^{2x-6} = 9$

Solution: Again, start by writing $3^{2x-6} = 3^2$, then use the property to say that $2x - 6 = 2$. Now solve this equation:

$$2x - 6 = 2$$
$$2x = 8$$
$$x = 4$$

11.3 Logarithms

> **What's Important**
>
> - Graph a logarithmic function.
> - Solve simple logarithmic equations.

If you look at the graphs of the exponential functions in the last section and try the horizontal line test, you'll see that they're both one-to-one. So they must have inverses. The inverse of an exponential function (with base b) is a **logarithmic function** (with base b), denoted $\log_b x$. The usual definition is this:

$$y = \log_b x \text{ if and only if } x = b^y$$

Note that since a positive number raised to a power is always positive, this means that you can only use positive numbers in logarithms.

Logarithms are actually all about finding exponents. For instance, if you have $\log_3 9$, you're really asking, "3 raised to what power equals 9?" The answer, of course, is 2. So in logarithmic form, we would write

$$\log_3 9 = 2$$

Graphing a logarithmic function is most easily done by making a table of values and plotting points (though you could take an exponential graph and reflect it across the line $y = x$). Just as there are two kinds of exponential graphs, there are two kinds of logarithmic graphs.

Example: Graph $f(x) = \log_2 x$

Solution: Make a table of values and plot points.

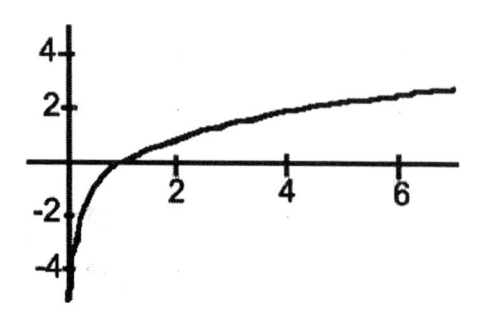

x	1/4	1/2	1	2	4
$f(x)$	-2	-1	0	1	2

11.3. LOGARITHMS

Example: Graph $f(x) = \log_{1/2} x$

Solution: Again, make a table of values and plot points.

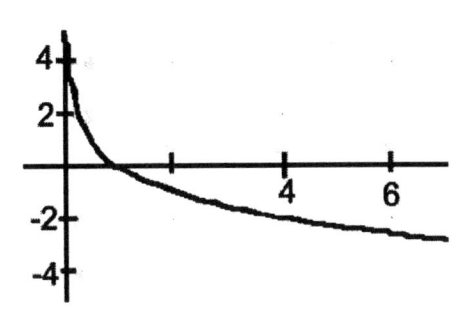

x	1/4	1/2	1	2	4
$f(x)$	2	1	0	-1	-2

By switching from logarithmic form to exponential form (using the definition of logarithms), we can solve some simple logarithmic equations.

Example: Solve $\log_3(x+1) = 4$

Solution: Use the definition to rewrite it in exponential form, like this:

$$x + 1 = 3^4$$

Then solve this equation.

$$x + 1 = 3^4$$
$$x + 1 = 81$$
$$x = 80$$

With logarithmic equations, you should always check your answer in the original equation to make sure that you're not taking the logarithm of zero or a negative number. In this case, 80 checks out.

Example: Solve $\log_b 9 = 1/2$

Solution: Again, start by changing to exponential form:

$$9 = b^{1/2}$$

Then solve this equation by squaring both sides. (Remember, $b^{1/2} = \sqrt{b}$.)

$$9 = \sqrt{b}$$
$$9^2 = (\sqrt{b})^2$$
$$81 = b$$

11.4 Properties of Logarithms

> **What's Important**
>
> - Use the properties of logarithms to work with logarithmic expressions.
> - Use a calculator to evaluate logarithms.

Logarithms have a bunch of properties which make them useful.

1. $\log_b b = 1$

2. $\log_b 1 = 0$

3. $\log_b b^x = x$

4. $b^{\log_b x} = x$

5. $\log_b MN = \log_b M + \log_b N$

6. $\log_b \dfrac{M}{N} = \log_b M - \log_b N$

7. $\log_b M^r = r \log_b M$

8. $\log_b M = \dfrac{\log_a M}{\log_a b}$ (change-of-base formula)

Properties 3 and 4 are based on the fact that exponentials and logarithms are inverses. Properties 5, 6 and 7 actually come from the properties of exponents (since logarithms are exponents). Learn these three properties! You'll use them all of the time when working with logarithms. Here are some examples of using them:

11.4. PROPERTIES OF LOGARITHMS

Example: Write $\log_3 \frac{xy}{z^2}$ as a combination of simple logarithms.

Solution: Just use the properties:

$$\begin{aligned}
\log_3 \frac{xy}{z^2} &= \log_3 xy - \log_3 z^2 &\text{(Property 6)} \\
&= \log_3 x + \log_3 y - \log_3 z^2 &\text{(Property 5)} \\
&= \log_3 x + \log_3 y - 2\log_3 z &\text{(Property 7)}
\end{aligned}$$

Example: Write $4\log_3 x - 5\log_3 y - 7\log_3 z$ as a single logarithm.

Solution: We'll again use properties 5, 6 and 7, but this time we'll read them right-to left.

$$\begin{aligned}
4\log_3 x - 5\log_3 y - 7\log_3 z &= \log_3 x^4 - \log_3 y^5 - \log_3 z^7 &\text{(Property 7)} \\
&= \log_3 x^4 - (\log_3 y^5 + \log_3 z^7) \\
&= \log_3 x^4 - \log_3 y^5 z^7 &\text{(Property 5)} \\
&= \log_3 \frac{x^4}{y^5 z^7} &\text{(Property 6)}
\end{aligned}$$

Example: Suppose $\log_b 2 = 0.356$ and $\log_b 5 = 0.827$. Find $\log_b 20$.

Solution: To find the value, we're going to use the properties to write $\log_b 20$ as a combination of $\log_b 2$ and $\log_b 5$. Start by factoring 20: $20 = 4 \cdot 5 = 2^2 \cdot 5$. Then we can say

$$\log_b 20 = \log_b 2^2 \cdot 5 = \log_b 2^2 + \log_b 5 = 2\log_b 2 + \log_b 5$$
$$= 2(0.356) + 0.827 = 0.712 + 0.827 = 1.539$$

Scientific calculators know how to do two kinds of logarithms:

- *Common logarithms* – logarithms with base 10, usually written $\log x$ (no number shown for the base). The calculator key sometimes just shows $\boxed{\log}$ or $\boxed{\text{LOG}}$.

- *Natural logarithms* – logarithms with base $e \approx 2.718281828$, usually written $\ln x$. The calculator key sometimes just shows $\boxed{\ln}$ or $\boxed{\text{LN}}$.

You may need to consult your calculator's manual on how exactly to use these buttons to find the logarithms. Remember that for most numbers, your calculator will only give an approximation of the actual value.

Example: Use a calculator to find $\log 15$ and $\ln 15$.

Solution: According to my trusty TI-30X,

$$\log 15 \approx 1.176091259$$

and

$$\ln 15 \approx 2.708050201$$

What if we wanted to find $\log_7 15$? 15 is not related to 7 (the base of the logarithm) in any natural way, so we can't calculate it directly, and if you look at your calculator, it doesn't have a key for $\log_7$. The answer to this question is Property 8 from above, the **change-of-base formula**. Here's how we would use it: The logarithm on the left is the one you're trying to find, so in this case $b = 7$. The logarithms on the right are in a base you (or your calculator) can find, so we can have $a = 10$ (for common logs) or $a = e$ (for natural logs). So

$$\log_7 15 = \frac{\log 15}{\log 7} = \frac{1.176091259}{0.84509804} = 1.391662509$$

Note that we would get the same result with natural logs:

$$\log_7 15 = \frac{\ln 15}{\ln 7} = \frac{2.708050201}{1.945910149} = 1.391662509$$

11.5 Logarithmic and Exponential Equations

What's Important

- Solve logarithmic equations.
- Solve exponential equations.

We've looked at some simple exponential and logarithmic equations back in Sections 11.2 and 11.3. In this section, we'll use the properties of logarithms to solve more complicated ones.

11.5. LOGARITHMIC AND EXPONENTIAL EQUATIONS

Example: Solve $\log_2 x + \log_2(x-1) = 1$

Solution: Start by using Property 5 to combine the logarithms on the left side together:

$$\log_2 x + \log_2(x-1) = \log_2 x(x-1)$$

Now we have an equation like the ones we solved in Section 11.3. Remember that we solve this by using the definition of logarithms to change to exponential form:

$$\log_2 x(x-1) = 1$$

so

$$x(x-1) = 2^1$$

And now just solve this equation.

$$x(x-1) = 2^1$$
$$x^2 - x = 2$$
$$x^2 - x - 2 = 0$$
$$(x-2)(x+1) = 0$$

So $x = 2$ or $x = -1$. Remember that we need to check our answers in the original equation. $x = 2$ doesn't cause problems, but notice that with $x = -1$ we'll have $\log_2(-1) + \log_2(-2)$, so we'll be taking logs of negative numbers, which isn't allowed. So throw out $x = -1$, so the only solution is $x = 2$.

Example: Solve $\log(x+1) - \log(x-1) = 4$

Solution: We'll solve this one the same way as the last example, by combining the logs on the left side, then converting to exponential form. Note that we have common logs here (log base

10).

$$\log(x+1) - \log(x-1) = 4$$
$$\log \frac{x+1}{x-1} = 4$$
$$\frac{x+1}{x-1} = 10^4$$
$$\frac{x+1}{x-1} = 10000$$
$$x+1 = 10000(x-1)$$
$$x+1 = 10000x - 10000$$
$$-9999x = -10001$$
$$x = \frac{10001}{9999}$$

Check your answer in the original equation and you find it doesn't cause problems.

Some logarithmic equations use an additional property of logarithms:

$$\text{If } \log_b M = \log_b N, \text{ then } M = N.$$

Example: Solve $\log x + \log(x+2) = \log 8$

Solution: For this problem, we're going to start by combining together the logs on the left side so that we end up with log = log. Then we can apply the property just given to eliminate the logs.

$$\log x + \log(x+2) = \log 8$$
$$\log x(x+2) = \log 8$$
$$x(x+2) = 8$$
$$x^2 + 2x = 8$$
$$x^2 + 2x - 8 = 0$$
$$(x+4)(x-2) = 0$$

So $x = -4$ or $x = 2$. As always, test these in the original equation. $x = -4$ causes problems (logs of negative numbers), but $x = 2$ works.

11.5. LOGARITHMIC AND EXPONENTIAL EQUATIONS

Exponential equations are usually solved by using Property 7 to bring down the exponent. Either common logs or natural logs are usually used.

Example: Solve $2^x = 7$

Solution: Start by taking the natural log of both sides, then bring down the exponent and solve for x:

$$2^x = 7$$
$$\ln 2^x = \ln 7$$
$$x \ln 2 = \ln 7$$
$$x = \frac{\ln 7}{\ln 2}$$
$$\approx 0.1771$$

Example: Solve $2^{x+1} = 7^{x-1}$

Solution: As in the last example, start by taking the natural log of both sides, then bring down the exponents and solve for x. Lot's more algebra in this one, though.

$$2^{x+1} = 7^{x-1}$$
$$\ln 2^{x+1} = \ln 7^{x-1}$$
$$(x+1) \ln 2 = (x-1) \ln 7$$
$$x \ln 2 + \ln 2 = x \ln 7 - \ln 7$$
$$x \ln 2 - x \ln 7 = -\ln 7 - \ln 2$$
$$x(\ln 2 - \ln 7) = -\ln 7 - \ln 2$$
$$x = \frac{-\ln 7 - \ln 2}{\ln 2 - \ln 7}$$
$$\approx 1.8671$$